Any con neurosciences should now about this great man; I really hope it inspires you as has done with me.

Best wishes

Cai.

January 2003

Advice for a Young Investigator

Advice for a Young Investigator

Santiago Ramón y Cajal

Translated by
Neely Swanson and
Larry W. Swanson

A Bradford Book
The MIT Press
Cambridge, Massachusetts
London, England

© 1999 Massachusetts Institute of Technology

This work originally appeared in Spanish under the title *Reglas y Consejos sobre Investigación Científica: Los tónicos de la voluntad.*

All rights reserved. No part of this book may be reproduced in any form by any electronic or mechanical means (including photocopying, recording, or information storage and retrieval) without permission in writing from the publisher.

This book was set in Palatino by Wellington Graphics and was printed and bound in the United States of America.

Library of Congress Cataloging-in-Publication Data

Ramón y Cajal, Santiago, 1852–1934
  [Reglas y consejos sobre investigacion cientifica. English]
  Advice for a young investigator / S. Ramón y Cajal ; translated by Neely Swanson and Larry W. Swanson.
  p. cm.
  "A Bradford book."
  Includes bibliographical references and index.
  ISBN 0-262-18191-6 (alk. paper)
  1. Research. 2. Scientists. I. Title.
Q180.A1R33313 1999
507.2—dc21   98-26036
                                                                          CIP

10 9 8 7 6 5 4

Contents

Foreword   vii
Preface to the second edition   xiii
Preface to the third edition   xvii
Preface to the fourth edition   xix

1. Introduction   1

Thoughts about general methods. Abstract rules are sterile. Need to enlighten the mind and strengthen resolve. Organization of the book

2. Beginner's Traps   9

Undue admiration of authority. The most important problems are already solved. Preoccupation with applied science. Perceived lack of ability

3. Intellectual Qualities   29

Independent judgment. Concentration. Passion for reputation. Patriotism. Taste for scientific originality

4. **What Newcomers to Biological Research Should Know**   53

   General education. The need for specialization. Foreign languages. How monographs should be read. The absolute necessity of seeking inspiration in nature. Mastery of technique. In search of original data

5. **Diseases of the Will**   75

   Contemplators. Bibliophiles and polyglots. Megalomaniacs. Instrument addicts. Misfits. Theorists

6. **Social Factors Beneficial to Scientific Work**   89

   Material support. Having a profession and doing research work are compatible. The investigator and his family

7. **Stages of Scientific Research**   111

   Observation. Experimentation. Working hypotheses. Proof

8. **On Writing Scientific Papers**   125

   Justification for scientific contributions. Bibliography. Justice and courtesy in decisions. Description of methods. Conclusions. The need for illustrations. Style. The publication of scientific works

9. **The Investigator as Teacher**   137

Foreword

Santiago Ramón y Cajal (1852–1934) is one of the more fascinating personalities in science. Above all he was the most important neuroanatomist since Andreas Vesalius, the Renaissance founder of modern biology. However, Cajal was also a thoughtful and inspired teacher, he made several lasting contributions to Spanish literature (his autobiography, a popular book of aphorisms, and reflections on old age), and he wrote one of the early books on the theory and practice of color photography. Furthermore, he was an exceptional artist, perhaps the best ever to draw the circuits of the brain, which he could never photograph to his satisfaction.

In his early thirties, Cajal wrote and illustrated the first original textbook of histology in Spain, which remained a standard throughout his lifetime. The first draft of his unique book of practical, fatherly advice to young people in the early stages of their research careers was begun soon after moving to the chair of histology and pathological anatomy at the University of Madrid about a decade later—when he also wrote the first major review of his investigations with Camillo Golgi's silver chromate method: *New Ideas on the Structure of the Nervous System* (1894).

This succinct book redefined how brain circuits had been described. In it, Cajal presented histological evidence that the central nervous system is not a syncytium or reticulum of cells as commonly believed at the time. Instead, it consists of individual neurons that usually conduct information in just one direction. The information output of the neuron is down a single axon and its branches to terminal boutons that end on or near the input side of another neuron (its cell body and dendrites). Cajal had discovered the synapse and with fundamental insight went on to describe the organization of all the major neural systems in terms of chains of independent neurons and the concept of functional polarity (unidirectional information flow in circuits). He was the first to explain in modern terms the organization of reflex and voluntary control pathways to the motor system, and this conceptual advance was the structural foundation of Sir Charles Sherrington's modern physiological explanation of reflexes and their control.

By the time the *Advice for a Young Investigator* was finally published, he was beginning to synthesize the vast research that established his reputation in a three volume masterpiece, the *Histology of the Nervous System in Man and Vertebrates* (1899–1904). So the *Advice* became a popular vehicle for Cajal to write down the thoughts and anecdotes he would give to students and colleagues about how to make important original contributions in any branch of science, and it was so successful that the third edition is still in print (in Spanish).

Part of the *Advice* is based on an analysis of his own success, while the rest comes from a judicious selection of wisdom from other places and other people's lives. Nevertheless, it is obviously Cajal's analysis of his own scientific career. As such, it is deeply embedded in contemporary Spanish culture and in the childhood of a country doctor's son. Hard work, ambition, patience, humility, seriousness, and passion for work, family, and country were among the traits he considered essential. But above all, master technique and produce original data; all the rest will follow.

It is interesting to compare Cajal the writer and Cajal the scientist. As a distinguished author of advice, autobiography, and reflections on life, he displayed a complex mixture of the romantic, idealist, patriot, and realist. And a sense of humor is obvious in his delightful chapter here on diseases of the will, where stereotypes of eccentric scientists are diagnosed according to symptoms we have all seen, and their prognosis discussed. In stark contrast, his scientific publications are almost ruthlessly systematic, descriptive, and deductive. He once wrote that his account of nervous system structure was not based on the appearance of a nerve cell here and there, but on the analysis of millions of neurons.

Because Cajal revealed so much about his thoughts and feelings in the *Advice* and in his autobiography, *Reflections on My Life*, it is easy to see his genius as well as his flaws. He deals with many broad issues of morals, religion, and patriotism that are often avoided, invariably generate controversy, and go in and out of fashion. However, it is important to bear in mind that he was writing in the late nineteenth century to aspiring researchers in his native Spain, which at the time was not one of the scientifically and politically elite countries of Europe. Thus, some of his advice may now appear dated or irrelevant to young people in North America and Europe who enjoy relative peace, prosperity, and intellectual security. However, it may become relevant to them sometime in the future, and it still applies to many other cultures.

This translation is based on two sources, the fourth edition of *Reglas y Consejos sobre Investigación Biológica (los tonicos de la voluntad)* (1916), and an English translation of the sixth edition by J.Mª. Sanchez-Perez, which was edited and annotated by C.B. Courville as *Precepts and Counsels in Scientific Investigation. Stimulants of the Spirit* (1951). We had originally thought that it would be worthwhile simply to reprint the Sanchez-Perez and Courville work, but finally decided that the translation was too literal, and in some few cases inaccurate, for today's students. Our goal has been to write a modern rather than literal translation, retaining as much flavor of the original as we could. The fourth edition was published when Cajal was over sixty, and was never substantially revised again. The later Spanish editions have two chapters at the end that are concerned primarily with conditions in Spain at the time, and they have not been translated because of their limited relevance today. We thank Graciela Sanchez-Watts for help with translating certain difficult passages.

Larry W. Swanson
Los Angeles, February 1, 1998

Selected Reading

Albarracín, A. (1982) *Santiago Ramón y Cajal o la Pasión de España*, with an introduction by P. L. Entralgo. Barcelona: Labor. An excellent collection of photographs and information about Cajal.

Cajal, S. Ramón y (1894) *Les nouvelles idées sur la structure du système nerveux chez l'homme et chez les vertébrés*, translated by L. Azoulay, with a prologue by M. Duval. Paris: Reinwald et Cie. For English translation by N. Swanson and L. W. Swanson, see *New Ideas on the Structure of the Nervous System in Man and Vertebrates*. Cambridge, MA: The MIT Press.

Cajal, S. Ramón y (1897) *Fundamentos Racionales y Condiciones Téchnicas de la Investigación Biológica*. Discurso leído ante la Real Academia de Ciencias Exactas, Físicas y Naturales en la recepción del Sr. D. Santiago Ramón y Cajal el día 5 de diciembre de 1897. Madrid: L. Aguado. The first appearance of the *Advice* in print. The title *Reglas y Consejos sobre Investigación Biológica* was used in the second edition (Madrid: Fortanet, 1899), and *(los tonicos de la voluntad)* was added to the fourth edition (Madrid: Fortanet, 1916).

Cajal, S. Ramón y (1909, 1911) *Histologie du système nerveux de l'homme et des vertébrés*, 2 vols., translated by L. Azoulay. Paris: Maloine. For English translation by N. Swanson and L. W. Swanson see *Histology of the Nervous System of Man and Vertebrates*, 2 vols. New York: Oxford University Press, 1995.

Cajal, S. Ramón y (1916) *Reglas y Consejos sobre Investigación Biológica (los tonicos de la voluntad)*, fourth edition. Madrid: Fortanet.

Cajal, S. Ramón y (1951) *Precepts and Counsels in Scientific Investigation. Stimulants of the Spirit*, translated by J. Mª. Sanchez-Perez, edited and annotated by C. B. Courville. Mountain View, CA: Pacific Press Publishing Association.

[Cajal, S. Ramón y] (1978) *Ramon y Cajal, 1852–1934, Expedientes administrativos de grandes Españoses*, 2 vols. Madrid: Ministerio de Educacion y Ciencia. An incredible collection of Cajal memorabilia.

Cajal, S. Ramón y (1989) *Recollections of My Life*, translated by E. H. Craigie with the assistance of J. Cano. Cambridge, MA: The MIT Press. A classic of scientific autobiography.

Cajal, S. Ramón y (1995) *Reglas y Consejos sobre Investigación Biológica (los tonicos de la voluntad)*, third edition, thirteenth Espasa Calpe printing, with a prologue by S. Ochoa. Madrid: Espasa Calpe.

Craigie, E. H. and W. C. Gibson (1968) *The World of Ramón y Cajal, with Selections from His Nonscientific Writings*. Springfield, IL: C. C Thomas.

DeFelipe, J., and E. G. Jones (1988) *Cajal on the Cerebral Cortex*. New York: Oxford University Press.

DeFelipe, J., and E. G. Jones (1992) Santiago Ramón y Cajal and methods in neurohistology. *Trends in Neurosciences*, 15:237–246.

Romero, A. (1984) *Fotografia Aragonesa/1: Ramon y Cajal*. Zaragoza: Diputación Provincial de Zaragoza. A wonderful collection of rare photographs.

Preface to the Second Edition

Paid for through the generosity of Dr. Lluria

This booklet is a reproduction, with several improvements and elaborations, of my speech on the occasion of being inducted into the Academia de Ciencias Exactas, Físicas y Naturales (at the meeting on December 5, 1897).

As with many other academic speeches that are more deserving of attention than ours, it remained forgotten on the shelves of official libraries until our dear friend Dr. Lluria generously reprinted it at his own expense as a gift to students and others interested in science.

Dr. Lluria (may God bless him with beautiful visions) believed that the advice and observations contained in this work may, like the feelings associated with intense research, help promote in our youth a love and enthusiasm for laboratory study. However, I don't know whether the advice, which is expressed with such exaggerated passion, will have any positive effect on the training of investigators.

I say this only because I got no such advice from relatives or teachers when forming the reckless desire to devote myself to the religion of the laboratory. At the time, I was usually doing things that didn't work, was lost, and more than once was hopelessly discouraged about my ability to pursue laboratory science. However, there were successes, despite my ignorance of the literature (unfortunately, not only through lack of diligence but also through lack of funds) and an inability to find anyone who could guide me, to help me discover anatomical studies that were already published in languages that I didn't know but should have. And there were so many times—for lack of discipline and more than anything for living so far from the kind of scientific atmosphere that stimulates and energizes the young investigator—that I thought about abandoning my work, tired and disgusted (as much from the work as from my sad and exhausting isolation), until I began to see the first tenuous flashes of a new idea!

Scientific routine and mental servitude to foreign opinion reigned despotically in our schools when I, a humble physician and recent graduate of the lecture halls with no prestigious official credentials, announced that I would publish an experimental study on *inflammation* (as any youthful study, it was poor and incomplete, but demonstrated good intentions and a love of research). One of the professors at my beloved University of Zaragoza (and certainly not one of the worst) was stupefied: "Who is Cajal to dare challenge the work of scholars!" At the time (1880), this professor was the university publicist and one of its most modern and open-minded thinkers. However, he was defending the belief (which, unfortunately, is still professed by so many of our intellectuals, who are genuinely ignorant or are simply justifying their own laziness under the guise of expediency) that scientific discoveries are not the fruit of methodical labor, but instead are gifts from God—gifts generously bestowed by Providence on a few priviledged souls invariably belonging to the hardest working nations, in other words France, England, Germany, and Italy.

Times have changed completely. Today, the investigator in Spain no longer works in isolation. While not large, there is nevertheless a group of young enthusiasts who stay in constant communication about their ideas and feelings because of their love for science and desire to collaborate on the magnum opus of progress. Today, at last, the questions that all aficionados of science ask when taking the first uncertain steps along the path of research have lost their desperate inhibitory grasp: Who cares what I do? To whom will I confide the joy that my small discovery has given me, someone who will not smile sarcastically or enter the realm of annoying compassion? If I triumph, who will applaud? And if I am uncertain, who will correct me and provide the encouragement to go on?

Under the guise of friendly criticism, some readers of this discourse have warned me that I place far too much emphasis on the discipline of will power and not enough on the exceptional aptitudes of great investigators. I would be the last to deny that the greatest scientific pioneers belonged to an aristocracy of the spirit and were exceptionally intelligent, something that we as modest investigators will never attain, no matter how much we exert ourselves. Nevertheless, before making this concession—which is entirely justified—I continue to believe that there is always room for anyone with average intelligence and an eagerness for recognition to utilize his energy and tempt fate. Like the lottery, fate doesn't always smile on the rich; from time to time it brings joy to the homes of the lowly. Instead, consider the possibility that any man could, if he were so inclined, be the *sculptor of his own brain,* and that even the least gifted may, like the poorest land that has been well cultivated and fertilized, produce an abundant harvest.

I could be mistaken, but conversations with illustrious scholars during my travels abroad lead me to the sincere belief that most of them have normal intelligence, although it has been highly refined, and that they are driven by an ardent longing for fame. But more than that, on occasion I have met renowned scholars whose intelligence and emotional qualities are inferior to the very discoveries that raised them from obscurity—discoveries that were made through the blind and unexpected workings of chance. The example of Courtois is more common than many realize; one ingenious writer has pointed out that *it is unclear whether he discovered iodine or iodine discovered him.*

In any event, what does it cost to prove whether or not we have the ability to create original science? In the end, how can we know whether there is a gift of exceptional aptitude for science among us if we don't try to create the opportunity for it to be expressed, under the influence of first-rate moral and technical discipline? As Balmes pointed out, "If Hercules had never tried to handle anything more than a stick, he never would have had the confidence to brandish a heavy club."

I hope that this modest booklet, which is meant for studious young people, will serve to increase their love for laboratory work; I also hope that it will rekindle the aspirations of those who believe in our intellectual and scientific rebirth, which have fallen somewhat in the light of serious recent misfortunes!

Madrid, December 20, 1898

Preface to the Third Edition

The edition generously paid for by Dr. Lluria has been out of print for more than three years. To satisfy American demand, we were obliged to reprint this booklet in two of their scientific journals. We also granted the same license to a Spanish literary/scientific corporation when we realized that abandoning this irrelevant little book would be a sin of negligence, as well as disadvantageous.

We have no illusions about the merit of our *discourse:* it has major defects from both the philosophical and literary points of view. Nevertheless, we have undoubtedly managed to enrich and improve the text, along with correcting several flaws, by a more thorough and selective reading of the philosophical and pedagogical literature, and by the experience gained during an additional fifteen years of teaching. But more than anything, the *postscript*—which was added to the end of the second edition at Dr. Lluria's request—has been completely rewritten. This was a wrenching task because an anguished heart cannot control the rhythm of its beat. It would have been inopportune and indiscreet under the circumstances to reprint it now as it appeared then.

However, we decided not to undertake a detailed editing of this modest little product of youth. Whether good or bad, every book has a spiritual personality. The public knows this and demands that the author respect it; they do not want it replaced under the guise of improvement. And this could very easily happen today, when, on the threshold of old age, we appear (and occasionally are) somehow defective. It is precisely this feature that attracts the reader's attention and gains his sympathy—just as with men, we admire and respect books for their good qualities; but we can only love them for certain faults that they display.

So, in the event that this feeling is not an illusion, we preserved essentially the same flavor in this third edition of the 1897 text. We made relatively few alterations in style and added only a paragraph here and there to help explain certain ideas that were highlighted superficially in the text. In addition, however, the present edition contains several new chapters, including the final one where we outline to the best of our humble understanding the work that the Spanish educational institutions (especially the *Junta de Pensiones y Ampliación de estudion en el extranjero*) have been charged with accomplishing—to assure that our country may collaborate in the management of universal culture and civilization in the shortest possible time, based on our intellectual and financial resources.

Madrid, January 1912

Preface to the Fourth Edition

I feel that urgent business and increased constraints on time have retarded the growth and improvement of the text of this booklet, which has been out of print for more than a year thanks to its increasing popularity with our generous intellectual young people. After dignified and courteous correspondence, especially with foreign scholars eager to translate this work, I had intended to universalize the text by purging it as much as possible of certain lively pleas and patriotic effusions that sound out of place or strident to the ear of citizens in those lucky nations where science is fortified by tradition, and students do not have to expend as much energy because they are already looked after by a fervent and self-sacrificing group of scholars. But, I repeat, forces stronger than my desires have spurned my impulse to revise. I wrote this book for Spain, and for now it should remain among Spaniards.

Despite the haste with which this fourth edition was revised, I have as usual introduced several modifications that seemed useful to me. I removed what seemed overly florid or ill-conceived; I honed certain passages where the style was tired, thick, or incorrect; and finally, I further developed several chapters by enriching them with new examples or pertinent observations.

I sincerely believe that this edition, more than the others, merits the unusual attention it has received from the new generation, along with the endorsement of certain illustrious scientists, for whose good will I am grateful.

Madrid, December 6, 1916

Advice for a Young Investigator

1   Introduction

Thoughts about general methods. Abstract rules are sterile. Need to enlighten the mind and strengthen resolve. Organization of the book

I shall assume that the reader's general education and background in philosophy are sufficient to understand that the major sources of knowledge include observation, experiment, and reasoning by induction and deduction.

Instead of elaborating on accepted principles, let us simply point out that for the last hundred years the natural sciences have abandoned completely the Aristotelian principles of intuition, inspiration, and dogmatism.

The unique method of reflection indulged in by the Pythagoreans and followers of Plato (and pursued in modern times by Descartes, Fichte, Krause, Hegel, and more recently at least partly by Bergson) involves exploring one's own mind or soul to discover universal laws and solutions to the great secrets of life. Today this approach can only generate feelings of sorrow and compassion—the latter because of talent wasted in the pursuit of chimeras, and the former because of all the time and work so pitifully squandered.

The history of civilization proves beyond doubt just how sterile the repeated attempts of metaphysics to guess at nature's laws have been. Instead, there is every reason to believe that when the human intellect ignores reality and concentrates within, it can no longer explain the simplest inner workings of life's machinery or of the world around us.

The intellect is presented with phenomena marching in review before the sensory organs. It can be truly useful and productive only when limiting itself to the modest tasks of observation, description, and comparison, and of classification that is based on analogies and differences. A knowledge of underlying causes and empirical laws will then come slowly through the use of inductive methods. Another commonplace worth repeating is that science cannot hope to solve Ultimate Causes. In other words, science can never understand the foundation hidden below the appearance of phenomena in the universe. As Claude Bernard has pointed out, researchers cannot transcend the determinism of phenomena; instead, their mission is limited to demonstrating the *how,* never the *why,* of observed changes. This is a modest goal in the eyes of philosophy, yet an imposing challenge in actual practice. Knowing the conditions under which a phenomenon occurs allows us to reproduce or eliminate it at will, therefore allowing us to control and use it for the benefit of humanity. Foresight and action are the advantages we obtain from a deterministic view of phenomena.

The severe constraints imposed by determinism may appear to limit philosophy in a rather arbitrary way.[1] However, there is no denying that in the natural sciences—and especially in biology—it is a very effective tool for avoiding the innate tendency to explain the universe as a whole in terms of general laws. They are are like a germ with all the necessary parts, just as a seed contains all the potentialities of the future tree within it. Now and then philosophers invade the field of biological sciences with these beguiling generalizations, which tend to be unproductive, purely verbal solutions lacking in substance. At best, they may prove useful when viewed simply as working hypotheses.

Thus, we are forced to concede that the "great enigmas" of the universe listed by Du Bois-Raymond are beyond our understanding at the present time. The great German physiologist pointed out that we must resign ourselves to the state of *ignoramus,* or even the inexorable *ignorabimus.*

There is no doubt that the human mind is fundamentally incapable of solving these formidable problems (the origin of life, nature of matter, origin of movement, and appearance of consciousness). Our brain is an organ of action that is directed toward practical tasks; it does not appear to have been built for discovering the ultimate causes of things, but rather for determining their immediate causes and invariant relationships. And whereas this may appear to be very little, it is in fact a great deal. Having been granted the immense advantage of participating in the unfolding of our world, and of modifying it to life's advantage, we may proceed quite nicely without knowing the essence of things.

It would not be wise in discussing general principles of research to overlook those panaceas of scientific method so highly recommended by Claude Bernard, which are to be found in Bacon's *Novum Organum* and Descartes's *Book of Methods.* They are exceptionally good at stimulating thought, but are much less effective in teaching one how to discover. After confessing that reading them may suggest a fruitful idea or two, I must further confess an inclination to share De Maistre's view of the *Novum Organum:* "Those who have made the greatest discoveries in science never read it, and Bacon himself failed to make a single discovery based on his own rules." Liebig appears even more harsh in his celebrated *Academic Discourse* when he states that Bacon was a scientific dilettante whose writings contain nothing of the processes leading to discovery, regardless of inflated praise from jurists, historians, and others far removed from science.

No one fails to use instinctively the following general principles of Descartes when approaching any difficult problem: "Do not acknowledge as true anything that is not obvious, divide a problem into as many parts as necessary to attack it in the best way, and start an analysis by examining the simplest and most easily understood parts before ascending gradually to an understanding of the most complex." The merit of the French philosopher is not based on his application of these principles but rather on having formulated them clearly and rigorously after having profited by them unconsciously, like everyone else, in his thinking about philosophy and geometry.

I believe that the slight advantage gained from reading such work, and in general any work concerned with philosophical methods of investigation, is based on the vague, general nature of the rules they express. In other words, when they are not simply empty formulas they become formal expressions of the mechanism of understanding used during the process of research. This mechanism acts unconsciously in every well-organized and cultivated mind, and when the philosopher reflexly formulates psychological principles, neither the author nor the reader can improve their respective abilities for scientific investigation. Those writing on logical methods impress me in the same way as would a speaker attempting to improve his eloquence by learning about brain speech centers, about voice mechanics, and about the distribution of nerves to the larynx—as if knowing these anatomical and physiological details would create organization where none exists, or refine what we already have.[2]

It is important to note that the most brilliant discoveries have not relied on a formal knowledge of logic. Instead, their discoverers have had an acute inner logic that generates ideas with the same unstudied unconsciousness that allowed Jourdain to create prose. Reading the work of the great scientific pioneers such as Galileo, Kepler, Newton, Lavoisier, Geoffroy Saint-Hilaire, Faraday, Ampere, Bernard, Pasteur, Virchow, and Liebig is considerably more effective. However, it is important to realize that if we lack even a spark of the splendid light that shone in those minds, and at least a trace of the noble zeal that motivated such distinguished individuals, this exercise may if nothing else convert us to enthusiastic or insightful commentators on their work—perhaps even to good scientific writers—but it will not create the spirit of investigation within us.

A knowledge of principles governing the historical unfolding of science also provides no great advantage in understanding the process of research. Herbert Spencer proposed that intellectual progress emerges from that which is homogeneous and that which is heterogeneous, and by virtue of the *instability of that which is homogeneous,* and of the principle that *every cause produces more than one effect,* each discovery immediately stimulates many other discoveries. However, even if this concept allows us to appreciate the historical march of science, it cannot provide us with the key to its revelations. The important thing is to discover how each investigator, in his own special domain, was able to segregate heterogeneous from homogeneous, and to learn why many of those who set out to accomplish a particular goal did not succeed.

Let me assert without further ado that there are no rules of logic for making discoveries, let alone for converting those lacking a natural talent for thinking logically into successful researchers. As for geniuses, it is well-known that they have difficulty bowing to rules—they prefer to make them instead. Condorcet has noted that "The mediocre can be educated; geniuses educate themselves."

Must we therefore abandon any attempt to instruct and educate about the process of scientific research? Shall we leave the beginner to his own devices, confused and abandoned, struggling without guidance or advice along a path strewn with difficulties and dangers?

Definitely not. In fact, just the opposite—we believe that by abandoning the ethereal realm of philosophical principles and abstract methods we can descend to the solid ground of experimental science, as well as to the sphere of ethical considerations involved in the process of inquiry. In taking this course, simple, genuinely useful advice for the novice can be found.

In my view, some advice about what should be known, about what technical education should be acquired, about the intense motivation needed to succeed, and about the carelessness and inclination toward bias that must be avoided, is far more useful than all the rules and warnings of theoretical logic. This is the justification for the present work, which contains those encouraging words and paternal admonitions that the writer would have liked so much to receive at the beginning of his own modest scientific career.

My remarks will not be of much value to those having had the good fortune to receive an education in the laboratory of a distinguished scientist, under the beneficial influence of living rules embodied in a learned personality who is inspired by the noble vocation of science combined with teaching. They will also be of little use to those energetic individuals—those gifted souls mentioned above—who obviously need only the guidance provided by study and reflection to gain an understanding of the truth. Nevertheless, it is perhaps worth repeating that they may prove comforting and useful to the large number of modest individuals with a retiring nature who, despite yearning for reputation, have not yet reaped the desired harvest, due either to a certain lack of determination or to misdirected efforts.

This advice is aimed more at the spirit than the intellect because I am convinced, and Payot wisely agrees, that the former is as amenable to education as the latter. Furthermore, I believe that all outstanding work, in art as well as in science, results from immense zeal applied to a great idea.

The present work is divided into nine chapters. In the second I will try to show how the prejudices and lax judgment that weaken the novice can be avoided. These problems destroy the self-confidence needed for any investigation to reach a happy conclusion. In the third chapter I will consider the moral values that should be displayed—which are like stimulants of the will. In the fourth chapter I will suggest what needs to be known in preparing for a competent struggle with nature. In the fifth, I will point out certain impairments of the will and of judgment that must be avoided. In the sixth, I will discuss social conditions that favor scientific work, as well as influences of the family circle. In the seventh, I will outline how to plan and carry out the investigation itself (based on observation, explanation or hypothesis, and proof). In the eighth I will deal with how to write scientific papers; and finally, in the ninth chapter the investigator's moral obligations as a teacher will be considered.

Notes

1. In attempting to prove his hypothesis, Claude Bernard may have exaggerated somewhat in claiming that: "We shall never know why opium has soporific effects, or why the combination of hydrogen and oxygen yields a substance so different in physical and chemical properties as water." The impossibility of reducing the properties of matter to laws governing the position, form, and movement of atoms (today we would say of ions and electrons) seems real at this time, but it does not seem that it should be thus in principle and forever. (Author's footnote, 1923.)

2. It is extraordinary how well this theory agrees with one elaborated by Schopenhauer (which was unknown to us at the time this essay was first published) in his book *The World as Will and as Representation,* pp. 98 ff. Concerning logic, he says that "the best logic for a particular science abandons the rules of logic when it begins serious discourse." And further on: "Wanting to make practical use of logic is like consulting the field of mechanics before learning to walk." More recently, Eucken expressed a similar view in saying that "rules and forms of logic are not enough to produce an ingenious thought." (Author's footnote, 1923.)

2   Beginner's Traps

Undue admiration of authority. The most important problems are already solved. Preoccupation with applied science. Perceived lack of ability

I believe that excessive admiration for the work of great minds is one of the most unfortunate preoccupations of intellectual youth—along with a conviction that certain problems cannot be attacked, let alone solved, because of one's relatively limited abilities.

Inordinate respect for genius is based on a commendable sense of fairness and modesty that is difficult to censure. However, when foremost in the mind of a novice, it cripples initiative and prevents the formulation of original work. Defect for defect, arrogance is preferable to diffidence, boldness measures its strengths and conquers or is conquered, and undue modesty flees from battle, condemned to shameful inactivity.

When one escapes the atmosphere of stylistic legerdemain inhaled while reading the published work of a genius, and enters the laboratory to confirm the observations upon which the intriguing ideas are based, now and then hero worship declines as self-esteem grows. Great men are at times geniuses, occasionally children, and always incomplete. Even when the work of a genius is subjected to critical analysis and no errors are found, it is important to realize that everything he has discovered in a particular field is almost nothing in comparison with what remains to be discovered. Nature offers inexhaustible wealth to all. There is certainly no reason to envy our predecessors, or to exclaim with Alexander following the victories of Philip, "My father is going to leave me nothing to conquer!"

Admittedly, certain concepts in science appear to be so complete, brilliant, and enduring that they seem to be the fruit of an almost divine intuition, springing forth perfect like Minerva from the head of Jupiter. However, the well-deserved admiration for such accomplishments would be considerably diminished were we aware of all of the time and effort, patience and perseverance, trials, corrections, and even mishaps that worked hand in hand to produce the final success—contributing almost as much as the investigator's genius. The same principle applies to the marvelous adaptation of the human organism to predetermined functions. When examined alone, the vertebrate eye or ear is a source of amazement. It seems impossible that these organs could have formed simply by the collective action of natural laws. However, when we consider all of the gradations and transitional forms that they display in the phylogenetic series, from the almost shapeless ocular outline of certain infusoria and worms to the complicated organization of the eye in lower vertebrates, not one whit of our admiration is lost and our minds are apt to accept the idea of natural formation through the mechanisms of variation, organic correlation, natural selection, and adaptation.[1]

What a wonderful stimulant it would be for the beginner if his instructor, instead of amazing and dismaying him with the sublimity of great past achievements, would reveal instead the origin of each scientific discovery, the series of errors and missteps that preceded it—information that, from a human perspective, is essential to an accurate explanation of the discovery. Skillful pedagogical tactics such as this would instill the conviction that the discoverer, along with being an illustrious person of great talent and resolve, was in the final analysis a human being just like everyone else.

Far from humbling one's self before the great authorities of science, those beginning research must understand that—by a cruel but inevitable law—their destiny is to grow a little at the expense of the great one's reputation. It is very common for those beginning their scientific explorations with some success to do so by weakening the pedestal of an historic or contemporary hero. By way of classic examples, recall Galileo refuting Aristotle's view of gravity, Copernicus tearing down Ptolemy's system of the universe, Lavoisier destroying Stahl's concept of phlogiston, and Virchow refuting the idea of spontaneous generation held by Schwann, Schleiden, and Robin. This principle is so general and compelling that it is displayed in all areas of science and extends to even the humblest of investigators. If I might be so bold as to refer to myself in the company of such eminent examples, I should add that on initiating my own work on the anatomy and physiology of nervous centers, the first obstacle that had to be set aside was the false theory of Gerlach and Golgi on the diffuse nature of neural networks in the gray matter, and on the nature of nerve current transmission.

Two phases may often be noted in the careers of learned investigators. First there is the productive time devoted to the elimination of past errors and the illumination of new data, and it is followed by the mature or intellectual phase (which does not necessarily coincide with old age) when scientific productivity declines and the hypotheses incubated during youth are defended with paternal affection from the attacks of newcomers.[2] Throughout history, no great man has shunned titles or failed to extol his right to glory before the new generation. Rousseau's bitter quote is sad but true: "There has never been a wise man who hasn't failed to prefer the lie invented by himself to the truth discovered by someone else."

Even in the most exact sciences there are always some laws that are maintained exclusively through the force of authority. To demonstrate their inaccuracy with new research is always an excellent way to begin genuine scientific work. It hardly matters whether the correction is received with harsh criticism, traitorous invective, or silence, which is even more cruel. Because right is on his side, the innovator will quickly attract the young, who obviously have no past to defend. And those impartial scholars who, in the midst of the smothering torrent of current doctrine, have learned how to keep their minds clear and their judgment independent will also gather on his side.

However, it is not enough to destroy—one must also build. Scientific criticism is justified only by establishing truth in place of error. Generally speaking, new principles emerge from the ruins of those abandoned, based strictly on facts correctly interpreted. The innovator must avoid all pious concessions to traditional error and crumbling ideas if he does not wish to see his fame quickly shared by the critics and those merely focusing on details, who immediately sprout in great numbers after each discovery, like mushrooms in the shade of a tree.

The Most Important Problems Are Already Solved

Here is another false concept often heard from the lips of the newly graduated: "Everything of major importance in the various areas of science has already been clarified. What difference does it make if I add some minor detail or gather up what is left in some field where more diligent observers have already collected the abundant, ripe grain. Science won't change its perspective because of my work, and my name will never emerge from obscurity."

This is often indolence masquerading as modesty. However, it is also expressed by worthy young men reflecting on the first pangs of dismay experienced when undertaking some major project. This superficial concept of science must be eradicated by the young investigator who does not wish to fail, hopelessly overcome by the struggle developing in his mind between the utilitarian suggestions that are part and parcel of his ethical environment (which may soon convert him to an ordinary and financially successful general practitioner), and those nobler impulses of duty and loyalty urging him on to achievement and honor.

Wanting to earn the trust placed in him by his mentors, the inexperienced observer hopes to discover a new lode at the earth's surface, where easy exploration will build his reputation quickly. Unfortunately, with his first excursions into the literature hardly begun, he is shocked to find that the metal lies deep within the ground—surface deposits have been virtually exhausted by observers fortunate enough to arrive earlier and exercise their simple right of eminent domain.

It is nevertheless true that if we arrived on the scene too late for certain problems, we were also born too early to help solve others. Within a century we shall come, by the natural course of events, to monopolize science, plunder its major assets, and harvest its vast fields of data.

Yet we must recognize that there are times when, on the heels of a chance discovery or the development of an important new technique, magnificent scientific discoveries occur one after another as if by spontaneous generation. This happened during the Renaissance when Descartes, Pascal, Galileo, Bacon, Boyle, Newton, our own Sanchez, and others revealed clearly the errors of the ancients and spread the belief that the Greeks, far from exhausting the field of science, had scarcely taken the first steps in understanding the universe.[3] It is a wonderful and fortunate thing for a scientist to be born during one of these great decisive moments in the history of ideas, when much of what has been done in the past is invalidated. Under these circumstances, it could not be easier to choose a fertile area of investigation.

However, let us not exaggerate the importance of such events. Instead, bear in mind that even in our own time science is often built on the ruins of theories once thought to be indestructible. It is important to realize that if certain areas of science appear to be quite mature, others are in the process of development, and yet others remain to be born. Especially in biology, where immense amounts of work have been carried out during the last century, the most essential problems remain unsolved—the origin of life, the problems of heredity and development, the structure and chemical composition of the cell, and so on.

It is fair to say that, in general, no problems have been exhausted; instead, men have been exhausted by the problems. Soil that appears impoverished to one researcher reveals its fertility to another. Fresh talent approaching the analysis of a problem without prejudice will always see new possibilities—some aspect not considered by those who believe that a subject is fully understood. Our knowledge is so fragmentary that unexpected findings appear in even the most fully explored topics. Who, a few short years ago, would have suspected that light and heat still held scientific secrets in reserve? Nevertheless, we now have *argon* in the atmosphere, the *x-rays* of Roentgen, and the *radium* of the Curies, all of which illustrate the inadequacy of our former methods, and the prematurity of our former syntheses.

The best application of the following beautiful dictum of Geoffroy Saint-Hilaire is in biology: "The infinite is always before us." And the same applies to Carnoy's no less graphic thought: "Science is a perpetual creative process." Not everyone is destined to venture into the forest and by sheer determination carve out a serviceable road. However, even the most humble among us can take advantage of the path opened by genius and by traveling along it extract one or another secret from the unknown.

If the beginner is willing to accept the role of gathering details that escaped the wise discoverer, he can be assured that those searching for minutiae eventually acquire an analytical sense so discriminating, and powers of observation so keen, that they are able to solve important problems successfully.

So many apparently trivial observations have led investigators with a thorough knowledge of methods to great scientific conquests! Furthermore, we must bear in mind that because science relentlessly differentiates, the minutiae of today often become important principles tomorrow.

It is also essential to remember that our appreciation of what is important and what is minor, what is great and what is small, is based on false wisdom, on a true anthropomorphic error. Superior and inferior do not exist in nature, nor do primary and secondary relationships. The hierarchies that our minds take pleasure in assigning to natural phenomena arise from the fact that instead of considering things individually, and how they are interrelated, we view them strictly from the perspective of their usefulness or the pleasure they give us. In the chain of life all links are equally valuable because all prove equally necessary. Things that we see from a distance or do not know how to evaluate are considered small. Even assuming the perspective of human egotism, think how many issues of profound importance to humanity lie within the protoplasm of the simplest microbe! Nothing seems more important in bacteriology than a knowledge of infectious bacteria, and nothing more secondary than the inoffensive microbes that grow abundantly in decomposing organic material. Nevertheless, if these humble fungi—whose mission is to return to the general circulation of matter those substances incorporated by the higher plants and animals—were to disappear, humans could not inhabit the planet.

The far-reaching importance of attention to detail in technical methodology is perhaps demonstrated more clearly in biology than in any other sphere. To cite but one example, recall that Koch, the great German bacteriologist, thought of adding a little alkali to a basic aniline dye, and this allowed him to stain and thus discover the tubercle bacillus—revealing the etiology of a disease that had until then remained uncontrolled by the wisdom of the most illustrious pathologists.

Even the most prominent of the great geniuses have demonstrated a lack of intellectual perspective in the appraisal of scientific insights. Today, we can find many seeds of great discoveries that were mentioned as curiosities of little importance in the writings of the ancients, and even in those of the wise men of the Renaissance. Lost in the pages of a confused theological treatise (*Christianismi restitutio*) are three apparently disdainful lines written by Servetus referring to the pulmonary circulation, which now constitute his major claim to fame. The Aragonese philosopher would be surprised indeed if he were to rise from the dead today. He would find his laborious metaphysical disquisitions totally forgotten, whereas the observation he used simply to argue for the residence of the soul in the blood is widely praised! Or again, it has been inferred from a passage of Seneca's that the ancients knew the magnifying powers of a crystal sphere filled with water. Who would have suspected that in this phenomenon of magnification, disregarded for centuries, slumbered the embryo of two powerful analytical instruments, the microscope and telescope—and two equally great sciences, biology and astronomy!

In summary, there are no small problems. Problems that appear small are large problems that are not understood. Instead of tiny details unworthy of the intellectual, we have men whose tiny intellects cannot rise to penetrate the infinitesimal. Nature is a harmonious mechanism where all parts, including those appearing to play a secondary role, cooperate in the functional whole. In contemplating this mechanism, shallow men arbitrarily divide its parts into essential and secondary, whereas the insightful thinker is content with classifying them as understood and poorly understood, ignoring for the moment their size and immediately useful properties. No one can predict their importance in the future.

Preoccupation with Applied Science

Another corruption of thought that is important to battle at all costs is the false distinction between *theoretical* and *applied* science, with accompanying praise of the latter and deprecation of the former. This error spreads unconsciously among the young, diverting them from the course of disinterested inquiry.

This lack of appreciation is definitely shared by the average citizen, often including lawyers, writers, industrialists, and unfortunately even distinguished statesmen, whose initiatives can have serious consequences for the cultural development of their nation.

They should avoid expressing the following sentiments: "Fewer doctors and more industrialists. The greatness of nations is not measured by what the former know, but rather by the number of scientific triumphs applied to commerce, industry, agriculture, medicine, and the military arts. We shall leave to the phlegmatic and lazy Teutons their subtle investigations of pure science and mad eagerness to pry into the remotest corners of life. Let us devote ourselves to extracting the practical essence of scientific knowledge, and then using it to improve the human condition. Spain needs machines for its trains and ships, practical advances for agriculture and industry, a rational health care system—in short, whatever contributes to the common good, the nation's wealth, and the people's well-being. May God deliver us from worthless scholars immersed in dubious speculation or dedicated to the conquest of the infinitesimal, which would be considered a frivolous if not ridiculous pastime if it weren't so expensive."

Ineptitudes like this are formulated at every step by those who, while traveling abroad, see progress as a strange mirage of effects rather than causes. People with little understanding fail to observe the mysterious threads that bind the factory to the laboratory, just as the stream is connected with its source. Like the man in the street, they believe in good faith that scholars may be divided into two groups—those who waste time speculating about unfruitful lines of pure science, and those who know how to find data that can be applied immediately to the advancement and comfort of life.[4]

Is it really necessary to dwell on such an absurd point of view? Does anyone lack the common sense to understand that applications derive immediately from the discovery of fundamental principles and new data? In Germany, France, and England the factory and laboratory are closely intertwined, and very often the scientist himself (either personally or through a development company) directs its industrial application. Such alliances are obvious in the great aniline dye factories that are one of the richest lodes of German, Swiss, and French industry. This is so well known that examples are hardly necessary. Nevertheless, I would like to cite two recent developments that are very significant. One is the great industry involved in the manufacture of precision lenses (for micrography, photography, and astronomy). It was created in Germany by the profound work in mathematical optics of Professor Abbé of Jena, and it gives Prussia an enormously valuable monopoly that is supported by the entire world.[5] The other example is the manufacture of therapeutic serums that was born in Berlin and perfected in Paris. It is both natural and legitimate that Behring and Roux, who established the scientific principles upon which serum therapy is based, exercise a controlling hand.

For the present, let us cultivate science for its own sake, without considering its applications. They will always come, whether in years or perhaps even in centuries. It matters very little whether scientific truth is used by our sons or by our grandsons. The course of progress obviously would have suffered if Galvani, Volta, Faraday, and Hertz, who discovered the fundamental principles of electricity, had discounted their findings because there were no industrial applications for them at the time.

Accept the view that nothing in nature is useless, even from the human point of view (with the necessary restrictions of time and place). Even in the rare instance where it may not be possible to use particular scientific breakthroughs for our comfort and benefit, there is one positive benefit—the noble satisfaction of our curiosity and the incomparable gratification and feeling of power that accompany the solving of a difficult problem.

In short, consider problems on their own merits when attacking them. Avoid deviating to secondary concerns that distract attention and weaken analytical powers. In struggling with nature, the biologist, like the astronomer, must look beyond the earth he lives on and concentrate on the serene universe of ideas, where the light of truth will eventually shine. The applications of new data will come in due time, when other related information emerges. It is well known that a discovery is simply the joining of two or more pieces of information to a useful end. Many scientific observations are of little use at the time they are made. However, after some decades, or perhaps even centuries, a new discovery clarifies the old, and the resulting *industrial application* may be called photography, the phonograph, spectral analysis, wireless telegraphy, or mechanical flight. Synthesis occurring over a variable length of time is always involved. Porta discovered the principle of the camera obscura, an isolated event that had very little impact on the art of design. Wedgwood and Davy noted in 1802 the possibility of obtaining photographic images on a certain type of paper immersed in silver nitrate solution, but this had little impact because the copy could not be fixed. Then came John Herschel, who succeeded in dissolving the silver salt not affected by light, and with this it was possible to fix the fugitive luminous silhouette. However, despite this advance, Porta's apparatus was virtually impossible to use because the silver salts available at the time were so weak. Then Daguerre finally appeared. He discovered the latent image in 1839 by using the much greater sensitivity of silver iodide. Daguerre admirably synthesized the inventions of his predecessors and used the foundation that they laid to create the science of photography as we know it today.

All inventions evolve in this way. Information is transmitted through time by discerning though unlucky observers who fail to harvest the fruits of their labor, which await fertilization. Nevertheless, once data are gathered, a scientist will come along at some point who is fortunate not so much for his originality as for having been born at an opportune moment. He considers the facts from the human point of view, synthesizes, and a discovery emerges.

Perceived Lack of Ability

Some people claim a lack of ability for science to justify failure and discouragement. "I enjoy laboratory work," they tell us, "but am no good at discovering things." Certainly there are minds unsuited for experimental work, especially if they have a short attention span and lack curiosity and admiration for the work of nature. But are the great majority of those professing incompetence really so? Might they exaggerate how difficult the task will be, and underestimate their own abilities? I believe that this is often the case, and would even venture to suggest that many people habitually confuse inability with the simple fact that they learn and understand slowly, or perhaps are sometimes even lazy or they don't have a secondary trait such as patience, thoroughness, or determination—which may be acquired rapidly through hard work and the satisfaction of success.

In my opinion the list of those suited for scientific work is much longer than generally thought, and contains more than the superior talents, readily adaptable, and keen minds ambitious for reputation and eager to link their names with a major discovery. The list also includes those ordinary intellects thought of as *skillful* because of the ability and steadiness they display for all manual work, those gifted with artistic talent who appreciate deeply the beauty of Nature's work, and those who are simply curious, calm, and phlegmatic devotees of the religion of detail, willing to dedicate long hours to examining the most insignificant natural phenomena. Science, like an army, needs generals as well as soldiers; plans are conceived by the former, but the latter actually conquer. Merely through being less brilliant, the collaboration of those who perfect and carry out the original plan cannot fail to be highly valuable. Thanks to these workers in the march of progress, the concept of a genius acquires vigor and clarity, transformed from abstract symbol to live reality, appreciated and known by all.

Various procedures can be used to assess one's aptitude for laboratory work. Based on my experience, I would recommend the following two:

1. Attempt to repeat some analytical method that is considered unreliable and difficult until patience and hard work yield results similar to those published by the author. Pleasure derived from success, especially if it has come without the supervision of an instructor (that is, working alone), is a clear indication of aptitude for experimental work.

2. Find a scientific topic that is difficult and surrounded by controversy, and examine it superficially by reading general reference books instead of detailed monographs. Then, after several months of experimental work, our beginner should consult the latest literature on the subject. If he has arrived at similar conclusions, if his thinking on hotly disputed points falls in line with the interpretations of noted authori-

ties, and if he has managed to avoid the errors committed by certain authors, then timidity should be abandoned, and scientific work should be approached without reservation. Many triumphs and satisfactions lie ahead, depending on how hard one works.

Even those with modest intellectual abilities will gather some fruit, provided they maintain faith in the creative power of education and devote extended periods of time to thorough analysis of a focused topic.

At the risk of appearing repetitive, tiresome, and boring, I would like to present the following reflections to counter those who do not believe in the power of determination. As many teachers and thinkers have noted, discoveries are not the fruit of outstanding talent, but rather of common sense enhanced and strengthened by technical education and a habit of thinking about scientific problems.[6] Thus, anyone with mental gifts balanced enough to cope with everyday life may use them to progress successfully along the road of investigation.

The youthful brain is wonderfully pliable and, stimulated by the *impulses of a strong will to do so,* can greatly improve its organization by creating new associations between ideas and by refining the powers of judgment.

Deficiencies of innate ability may be compensated for through persistent hard work and concentration. One might say that work substitutes for talent, or better yet that it *creates talent.* He who firmly determines to improve his capacity will do so, provided that education does not begin too late, during a period when the plasticity of nerve cells is greatly reduced. Do not forget that reading and thinking about masterpieces allows one to assimilate much of the skill that created them, providing of course that one extends beyond conclusions to the author's insights, guiding principles, and even style.

What we refer to as a great and special talent usually implies superiority that is *expeditious* rather than *qualitative*. In other words, it simply means doing quickly and with brilliant success what ordinary intellects carry out slowly but well. Instead of distinguishing between mediocre and great minds, it would be preferable and more correct in most instances to classify them as *slow* and *facile*.[7] The latter are certainly more brilliant and stimulating—there is no substitute for them in conversation, oratory, and journalism, that is, in all lines of work where time is a decisive factor. However, in scientific undertakings the *slow* prove to be as useful as the fast because scientists like artists are judged by the quality of what they produce, not by the speed of production. I would even venture to add that as a very common compensation *slow* brains have great endurance for prolonged concentration. They open wide, deep furrows in problems, whereas facile brains often tire quickly after scarcely clearing the land. There are, however, many exceptions to this generalization: Newton, Davy, Pasteur, Virchow, and others were active minds who left a broad, luminous wake.

If our memory is inconsistent and weak, despite efforts to improve, then let us *manage it well*. As Epictetus said: "When you are dealt poor cards in the game of life, there is nothing to do but make the best of them." History teaches of the occasional great discoveries made by those with ordinary minds and memory ably used, rather than by those with superior abilities. Great scientific innovators such as Helmholtz have complained of bad memory—of how learning prose by rote is akin to torture! As compensation, those with short memories for words and phrases seem to enjoy excellent retention of ideas and logical arguments. And Locke has pointed out that those endowed with great genius and a facile memory do not always excel in judgment.

To pursue fully the topic of our research with the limited facilities that we have, let us forget unrelated pursuits and the parasitic ideas connected with the futile trifles of everyday life. Using strength and perseverance, concentrate deeply only on information pertinent to the question at hand. During the gestation period of our work, sentence ourselves to ignorance of everything else that is going on—politics, literature, music, and idle gossip. There are occasions when ignorance is a great virtue, almost a state of heroism. Useless books distract attention and are thus weighty, occupying as much space in our brains as on the library shelf. They can spoil or hinder mental adjustments to the problem at hand. Although popular opinion may not agree, "Knowledge occupies space."

Even those with mediocre talent can produce notable work in the various sciences, so long as they do not try to embrace all of them at once. Instead, they should concentrate attention on one subject after another (that is, in different periods of time), although later work will undermine earlier attainments in the other spheres. This amounts to saying that the brain adapts to universal science in *time* but not in *space*. In fact, even those with great abilities proceed in this way. Thus, when we are astonished by someone with publications in different scientific fields, realize that each topic was explored during a specific period of time. Knowledge gained earlier certainly will not have disappeared from the mind of the author, but it will have become simplified by condensing into formulas or greatly abbreviated symbols. Thus, ample space remains for the perception and learning of new images on the cerebral blackboard.

Notes

1. I believe less in the power of natural selection today than I did when I wrote these lines in 1893. The more I study the organization of the eye in vertebrates and invertebrates, the less I understand the causes of their marvelous and exquisitely adapted organization.

2. Ostwald corroborates this view in a recent book, noting that almost all the great discoveries have been the work of youth. Newton, Davy, Faraday, Hertz, and Mayer are good examples.

3. The brilliant series of discoveries in electricity that followed Volta's development of the voltaic pile at the beginning of the last century, the Pleiades of histological work inspired by Schwann's discovery of cell multiplication, and the profound repercussions that the not so distant finding of roentgen rays have produced in all areas of physics (the observation of radioactivity, and the discovery of radium and polonium and of the phenomenon of emanation) are good examples of that creative and, in a sense, automatic virtue possessed by all great discoveries, which seem to grow and multiply like seeds cast by chance on fertile soil.

4. This popular view has been refuted eloquently by many scholars. However, I can't resist the temptation to quote a comparison that has been made in various brilliant forms, here by our incomparable scientific commentator, José Echegaray, who did so much to translate science into popular terms, and whose death robbed Spanish science of a great talent:

Pure science is like a beautiful cloud of gold and scarlet that diffuses wondrous hues and beams of light in the west. It is not an illusion, but the splendor and beauty of truth. However, now the cloud rises, the winds blow it over the fields, and it takes on darker, more somber colors. It is performing a task and changing its party clothes—think of it as putting on its work shirt. It generates rain that irrigates the fields, soaking the land and preparing it for future harvests. In the end it provides humanity with its daily bread. What began as beauty for the soul and intellect ends by providing nourishment for the humble life of the body. *Academia de Sciencias*, formal session of March 12, 1916.

5. This was written in 1896. Now [1923] there are no fewer than thirty-three outstanding researchers in mathematics, optics, mechanics, and chemistry

at the optical instruments factory in Jena. Furthermore, legions of chemists also work in the great German factories that produce chemical products. It is clear that the only way for industry to avoid routine and stagnation is to convert the laboratory to an antechamber of the factory.

6. "It is common sense to work under considerable stress," according to the graphic adage of Echegaray.

7. This view is consistent with the classification of *classic* and *romantic* (applied to minds that react slowly and minds that react quickly) provided by Ostwald in his interesting recent book, *Great Men*.

3 Intellectual Qualities

Independent judgment. Concentration. Passion for reputation. Patriotism. Taste for scientific originality

Indispensable qualities for the research worker include independent judgment, intellectual curiosity, perseverance, devotion to country, and a burning desire for reputation.

It is unnecessary to consider intellectual abilities in any detail. I assume that the newcomer to laboratory work is endowed with normal intelligence, a reasonable amount of imagination, and most of all, that harmonious coordination of faculties that is much more valuable than brilliant but erratic and unbalanced mental gifts.

Charles Richet has stated that the idealism of Don Quixote is combined with the good sense of Sancho in men of genius. The investigator should display some happy combination of these traits: an artistic temperament that impels him to search for and admire the number, beauty, and harmony of things; and—in the struggle for life that ideas create in our minds—a sound critical judgment that is able to reject the rash impulses of daydreams in favor of those thoughts most faithfully embracing objective reality.

Chapter 3

Independent Judgment

High-minded independence of judgment is a dominant trait shown by eminent investigators. They are not spellbound or overly impressed by the work of their predecessors and mentors but instead observe carefully and question. Geniuses such as Vesalius, Eustachius, and Harvey—who corrected the anatomical work of Galen—and others including Copernicus, Kepler, Newton, and Huygens, who overturned the ancient astronomy—were undoubtedly illustrious thinkers. Most importantly, however, they were ambitious and exacting individualists with extraordinarily bold critical insight. Saints may emerge from the docile and humble, but rarely scholars. I believe that excessive fondness for tradition, along with obstinate determination to maintain scientific formulations of the past, reflect either indomitable mental laziness or a blanket to cover mistakes.

Hapless is he who remains silent and absorbed in a book. Extreme admiration drains the personality and clouds understanding, which comes to accept hypothesis for proof and shadow for obvious truth.

I am sure that on first reading, not everyone is able to stumble across the gaps and flaws of an inspired book. Nevertheless, undue veneration, like all emotional states, prevents critical evaluation. If we feel spent after thought-provoking reading, allow a few days to pass. Then go on with a cool head and calm judgment to a second or even third reading. Little by little, deficiencies become apparent and flimsy logic is revealed. Ingenious hypotheses lose their authority, only to reveal their shaky foundations. We are no longer influenced by the magic of style. In short, understanding emerges. We are no longer a blind worshiper, but a judge, of the book. This is the moment for research to begin, for replacing the author's hypotheses with more reasonable ones, and for subjecting everything to intense criticism.

As with many beauties of nature, the enchantment of human works can only be retained when viewed from a distance. Analysis is the microscope that brings objects close to us and reveals the coarse weave of their tapestry. The illusion dissolves when the artificial nature of the embroidery and presence of design flaws become apparent to the eyes.

It could be said that in our times, when so many idols have been dethroned and so many illusions destroyed or forgotten, there is little need for resorting to a critical sense and spirit of doubt. Certainly they are not as necessary today as in times past. However, old habits die hard—too often one still encounters the pupils of illustrious men wasting their talents on defending the errors of their teachers, rather than using them to solve new problems. It is also important to note that in our era of disrespectful criticism and changing values, school discipline reigns with such tyranny in the universities of France, Germany, and Italy that at times even the greatest initiative is suffocated, along with the flowering of original thought. Those of us who battle alone like ordinary soldiers could cite many examples of servitude to this type of school discipline, and to political domination as well! We have known so many keen intellects who have had the misfortune of being pupils of great men! It is especially worth mentioning here those generous and grateful souls who know how to look for the truth, but dare not make it known for fear of snatching away from the master part of the prestige that will sooner or later fall to the thrust of less scrupulous adversaries because it is founded on error.

The mission of those docile souls who form the retinues of outstanding investigators—as susceptible to suggestion as they are to inactivity and perseverance in error—has always been to flatter genius and applaud its aberrations. This is the lip service mediocrity complacently renders to superior talent. It is easy to understand when one recalls that inferior intellect adapts better to error, which almost always involves a simple answer, than to truth, which is often rigorous and difficult.

Concentration

Those writing about logic emphasize with good reason the creative power of concentration, although they tend to ignore a variety that might appropriately be called *cerebral polarization* or *sustained concentration*—that is, steady orientation of all our faculties toward a single object of study for a period of months or even years. The thinking of countless brilliant minds ends up sterile for lack of this ability, which the French call *esprit de suite*. I could cite dozens of Spaniards with minds finely suited to scientific investigation who retreat discouraged from a problem without seriously measuring their strength, perhaps just at the moment when nature was about to reward their eagerness with the anxiously awaited revelation. Our classrooms and laboratories are full of these capricious and restless souls who love research and suffer through mishaps with the retort or microscope day after day. Their feverish activity yields an avalanche of lectures, articles, and books—upon which they have lavished a great deal of scholarship and talent. They constantly exhort the garrulous throng of dreamers and theorizers with the indispensable need for observing nature directly. Then, after long years of publicity and experimental work, those closest to them (their satellites at the prestigious yet mysterious meetings where the great preside) are asked about the discoveries of the master. The allies are forced to confess shamefacedly that the great burden of talent, combined with the virtual impossibility of summarizing in a nutshell the extraordinary magnitude and range of the work undertaken, make it impossible to state what partial or positive progress had been made. These are the inevitable fruits of negligence or excessive lack of focus, not to mention childish, encyclopedic ostentation. This approach is inconceivable today, when even the most renowned scholars specialize and concentrate in order to produce. But enough of this; we shall deal later with bad habits of the will.

To bring scientific investigation to a happy end once appropriate methods have been determined, we must hold firmly in mind the goal of the project. The object here is to focus the train of thought on more and more complex and accurate associations between images based on observation and ideas slumbering in the unconscious—ideas that only vigorous concentration of mental energy can raise to the conscious level. One must achieve total absorption; expectation and focused attention are not enough. We must take advantage of all lucid moments, whether they occur during the meditation following prolonged rest; during the superintense mental work nerve cells achieve when fired by concentration; or during scientific discussion, whose impact often generates unanticipated intuition like sparks from steel.

Most people who lack self-confidence are unaware of the marvelous power of prolonged concentration. This type of cerebral polarization (which involves a special ordering of perceptions) refines judgment, enriches analytical powers, spurs constructive imagination, and—by focusing all light of reason on the darkness of a problem—allows unforeseen and subtle relationships to be discovered. If a photographic plate under the center of a lens focused on the heavens is exposed for hours, it comes to reveal stars so far away that even the most powerful telescopes fail to reveal them to the naked eye. In a similar way, time and concentration allow the intellect to perceive a ray of light in the darkness of the most complex problem.

The comparison just made is not, however, entirely accurate. Photography in astronomy is limited to recording faint though preexisting stars, whereas intellectual work is an act of creation. It is as if the mental image that is studied over a period of time were to sprout appendages like an ameba—outgrowths that extend in all directions while avoiding one obstacle after another—before interdigitating with related ideas.

The forging of new truth almost always requires severe abstention and renunciation. During the so-called intellectual incubation period, the investigator should ignore everything unrelated to the problem of interest, like a somnambulist attending only to the voice of the hypnotist. In the lecture room, on walks, in the theater, in conversation, and even in reading for pleasure, seek opportunities for insight, comparisons, and hypotheses that add at least some clarity to the problem one is obsessed with. Nothing is useless during this process of adjustment. The first glaring errors, as well as the wrong turns ventured on by the imagination, are necessary because in the end they lead us down the correct path. They are part of the final success, just as the initial formless sketches of the artist are a part of the finished portrait.

When one reflects on the ability that humans display for modifying and refining mental activity related to a problem under serious examination, it is difficult to avoid concluding that the brain is plastic and goes through a process of anatomical and functional differentiation, adapting itself progressively to the problem. The adequate and specific organization acquired by nerve cells eventually produces what I would refer to as professional or adaptational talent. As a motivator of the will itself, this brain organization provides the energy to adapt understanding to the nature of the problem under consideration. In a certain sense, it would not be paradoxical to say that the person who initiates the solution to a problem is different from the one who solves it. This is an obvious and simple explanation for the astonishment proclaimed by all investigators on discovering the simple solution so laboriously sought. "Why didn't I think of this at the outset!" we exclaim. "There was so much confusion traveling down roads that led nowhere!"

If a solution fails to appear after all of this, and yet we feel success is just around the corner, try resting for a while. Several weeks of relaxation and quiet in the countryside brings calmness and clarity to the mind. Like the early morning frost, this intellectual refreshment withers the parasitic and nasty vegetation that smothers the good seed. Bursting forth at last is the flower of truth, whose calyx usually opens after a long and profound sleep at dawn—in those placid hours of the morning that Goethe and so many others consider especially favorable for discovery.

Travel has the same virtue of renewing thought and dissipating tiring preoccupations by furnishing new views of the world and transmitting our store of ideas to others. How often the powerful vibration of the locomotive and the spiritual solitude of the railway car (the "just rewards of humanity," as Descartes might say) suggest ideas that are ultimately confirmed in the laboratory!

Now that scientific research has become a regular profession on the payroll of the state, the observer can no longer afford to concentrate for extended periods of time on one subject, and must work even harder. Gone are the wonderful days of yore when those curious about nature were able to remain withdrawn in the silence of the study, confident that rivals would not disrupt their tranquil meditations. Research is now frantic. When a new technique is outlined, many scholars immediately take advantage of it and apply it almost simultaneously to the same problems—diminishing the glory of the originator, who probably lacks the facilities and time necessary to gather all the fruits of his labors, and of his lucky star.

As a result, the coincidences and battles of priority are inevitable. It is clear that once an idea becomes public it joins the intellectual atmosphere that nourishes all of our minds. Because of the functional synchronization that governs minds prepared and oriented toward a particular subject, the new idea is assimilated simultaneously in Paris and Berlin, in London and Vienna—in virtually the same way, with similar developments and applications. The discovery grows and develops spontaneously and automatically like an organism, as though scholars are reduced to mere cultivators of the seed planted by a genius. The magnificent flowering of new information is observed by all, and naturally everyone wishes to gather for themselves the splendid blossoms. This explains the eagerness to publish most laboratory studies, even when imperfect and incomplete. The desire to arrive first results at times in shallowness, although it is also true that feverish anxiety to reach the goal first wins the prize for priority.

Be that as it may, it is unwise to become disenchanted if someone arrives ahead of us. Continue work undaunted; in time our turn will come. That eminent woman, Madam Curie, provides an eloquent example of untiring perseverance. After discovering the radioactivity of thorium, she was unpleasantly surprised to learn that the same observation had been announced a short time earlier by Schmidt in the *Wiedermann Annalen*. Far from disheartened, however, she continued her research uninterrupted. She analyzed new substances with the electroscope, including uranium oxide (*pitchblende*) from the mines of Johann Georgenstadt, and its radioactivity proved four times stronger than that of uranium itself. Suspecting that this very active material contained a new element, she undertook (with the assistance of M. Curie) a series of ingenious, patient, and heroic experiments that were rewarded with the discovery of a new element, the remarkable radium. Its properties inspired a great deal of further work that has revolutionized chemistry and physics.

In Spain, where laziness is a religion rather than a vice, there is little appreciation for how the monumental work of German chemists, naturalists, and physicians is accomplished—especially when it would appear that the time required to execute the plan and assemble a bibliography might involve decades! Yet these books have been written in a year or two, quietly and without feverish haste. The secret lies in the method of work; in taking advantage of as much time as possible for the activity; in not retiring for the day until at least two or three hours are dedicated to the task; in wisely constructing a dike in front of the intellectual dispersion and waste of time required by social activity; and finally, in avoiding as much as possible the malicious gossip of the café and other entertainment—which squanders our nervous energy (sometimes even causing disgust) and draws us away from our main task with childish conceits and futile pursuits.

If our professions do not allow us to devote more than two hours a day to a subject, do not abandon the work on the pretext that we need four or six. As Payot wisely noted, "A little each day is enough, as long as a little is produced each day."

The harm in certain things that are too distracting lies not so much in the time they steal from us as in the enervation they bring to the creative tension of the mind, and in the loss they cause to that quality of tone that nerve cells acquire when adapted to a particular subject.

Of course we don't recommend the elimination of all distractions. However, those of the investigator should always be light and promote the association of new ideas. A stroll outside, contemplating works of art and photography, enjoying scenes such as monuments in different lands, the enchantment of music—and more than anything else the companionship of a person who understands us and carefully avoids all serious and reflective conversation—are the best ways for the laboratory worker to relax. Along these lines, it is wise to follow the advice of Buffon, who justified his abandon in conversation (which displeased many of those who admired the nobility, along with his elegant writing style) by noting: "These are my moments of rest."

In summary, all great work is the fruit of patience and perseverance, combined with tenacious concentration on a subject over a period of months or even years. Many illustrious scholars have confirmed this when questioned about the secret of their creations. Newton stated that he arrived at the sovereign law of universal attraction only by constant thinking about the same problem. According to one of his sons, Darwin achieved such a high degree of concentration on the biological facts related to the principle of evolution that for many years he systematically deprived himself of all reading and contemplation unrelated to the goal of his thoughts. Buffon said unreservedly, "Genius is simply patience carried to the extreme." To those who asked how he achieved fame he replied: "By spending forty years of my life bent over my writing desk." As a final example, it is widely known that Mayer, the genius who discovered the principle of energy conservation and transformation, dedicated his entire life to this concept.

Thus, it is clear beyond doubt that great scientific undertakings require intellectual vigor, as well as severe discipline of the will and continuous subordination of all one's mental powers to an object of study. Harm is caused unconsciously by the biographers of illustrious scholars when they attribute great scientific conquests to genius rather than to hard work and patience. What more could the weak will of the student or professor ask than to rationalize its laziness with the modest, and thus even more lamentable, admission of intellectual mediocrity! Not even biographers with the good sense of G.L. Figuier are immune from the regrettable trend of extolling beyond reason the mental gifts of famous investigators. Careful thought should make them realize how discouraging this can be to their readers. On the other hand, many autobiographies wherein the sage presents himself full-length to the reader provide an excellent moral tonic, showing weaknesses and passions, lapses and triumphs. After reading autobiographies that fill the soul with hope, you might well say: "Even I can be a painter!"

Passion for Reputation

The psychology of the researcher is somewhat different from the psychology of the typical intellectual. He is undoubtedly driven by the same aspirations and reacts to the same stimuli as the rest of humanity. But two emotions must be unusually strong in the great scientific scholar: a devotion to truth and a passion for reputation. The dominance of these two zeals explains the entire life of the investigator. The contrast between the ideal life he develops and the one that ordinary men hammer out produces the struggles, digressions, and misunderstandings that have always characterized relationships between scholar and social environment.

It has often been said that men of science, like great religious and other social reformers, exhibit mental traits frequently associated with the inept. They dwell on the high ground of humanity, unconcerned with the pettiness and trifles of everyday life.

Despite all this, the genuine scholar remains profoundly human, surpassing the best in his love for the rest of mankind. Extending in time and space, this feeling applies to friends and strangers alike, and is directed toward humanity now and in the future. Thanks to these unique talents—whose glance penetrates the obscurity of the future, and whose exquisite sensitivity demands regret for mistakes and for the stagnation of everyday life—social and scientific development is possible. Only the scholar is expected to fight the current, and in so doing alter the prevailing moral climate. It is important to repeat that his mission is not to adapt his ideas to those of society; instead, his mission is to adapt those of society to his own. And in the likely event that he is correct and proceeds with disciplined confidence and a minimum of conflict, sooner or later humanity will follow, applaud, and crown him with fame. Every investigator is driven by the hope of this promising tribute of veneration and justice because he knows that while individuals are capable of ingratitude, collective groups rarely are, once they are fully conscious of the truth and usefulness of an idea.

It is common knowledge that eagerness for approval and applause affects everyone to a greater or lesser extent—especially those endowed with a generous heart and clear understanding. Nevertheless, each of us seeks fame along a different path. Some proceed along the military lines celebrated by Cervantes in his *Don Quixote,* aspiring to advance the political might of their country. Others travel the road of art, eagerly seeking the ready applause of the masses, who understand beauty much better than truth. But very few in each country (though more in the most civilized) take the path of scientific investigation, the only path that can lead us to a rational and positive explanation of man and surrounding nature. I hold that this ambition is one of the worthiest and most laudable that man can pursue because perhaps more than any other it is infused with the fragrance of universal love and charity.

The contrast between the moral stance of scholars and patriots is a topic of frequent discussion. Because we live in a country that has sacrificed too much on the altar of its heroes (warriors, politicians, and members of religious orders) while forsaking (if not persecuting) its most original thinkers, may I be allowed to present a eulogy to the opposite point of view?

Heroes and scholars represent the opposite extremes of energy production and yet are equally necessary for the progress and well-being of humanity. However, the importance of their work is quite different. The scholar struggles for the benefit of all humanity, sometimes to reduce physical effort, sometimes to reduce pain, and sometimes to postpone death, or at least render it more bearable. In contrast, the patriot sacrifices a rather substantial part of humanity for the sake of his own prestige. His statue is always erected on a pedestal of ruins and corpses; his triumphs are celebrated exclusively by a tribe, party, or nation; and he leaves behind a wake of hatred and bloody waste in the conquered territory. In contrast, all humanity crowns a scholar, love forms the pedestal of his statue, and his triumphs defy the desecration of time and the judgment of history. His only victims (if those unsaved from ignorance can be regarded as such) are the laggards, the atavistic, and those who profit from lies and errors—in short, those who would be denounced as open enemies of happiness and the worthwhile in any well-organized society.

Fortunately, there is no shortage in our country of high minds who place their chances of happiness on winning the approbation of public opinion. But unfortunately, these talented men—with rare and commendable exceptions—prefer to win the laurel wreath by following the path of art or literature, where the great majority fail or do not attain the hoped-for recognition. With the exception of a few very eminent artistic and literary geniuses whose work is appreciated and praised abroad, how few of our painters and poets will be worshiped by posterity! How many who struggle in vain to create worldwide reputations as writers or orators could achieve it instead as investigators of science, perhaps with less effort! It is so difficult to be original in a field where virtually everything was drained by the ancients, who were endowed with a marvelous sense of literary beauty and plastic form, leaving hardly anything to be gleaned in the field of art!

After reading the orations of Demosthenes and Cicero, the dialogues of Plato, the parallel lives of Plutarch, and the speeches of Titus Livius, it is easy to be convinced that no modern orator has succeeded in devising a genuinely original method of persuading the intellect or moving the human heart. The role of the contemporary orator is to apply the innumerable points of form and argumentation thought out by the authors of the classics to specific and more or less novel situations.

And what are we to think about those who seek the prestige of originality in poetry and artistic prose? After Homer and Virgil, Horace and Seneca, Shakespeare and Milton, Cervantes and Ariosto, Goethe and Heine, Lamartine and Victor Hugo, Chateaubriand and Rousseau, and many others, what audacious soul would aspire to invent a poetic figure, a blend of expression full of emotion, or an exquisiteness of style that had not been expressed by those incomparable minds?

We don't deny that artistic creations equal to or even greater than those bequeathed to us by the classics are possible. The great monuments elected by the versatile writers of the Renaissance, and the sublime creations of the Romantic school during the last century, bear witness to the fact that the vein of original literature remains far from exhausted. We are simply pointing out that literary compositions of merit are extremely difficult to achieve, and that they extract more anxiety and labor than original scientific work. The reason is obvious. Because art depends on popular judgments about the universe, and is nourished by the limited expanse of sentiment, it has had time to drain virtually all of the emotional content from the human soul, beauties of the external world, and ingenious combinations of verbal images. In contrast, science was barely touched upon by the ancients, and is as free from the inconsistencies of fashion as it is from the fickle standards of taste. It accumulates new results day by day, providing endless work for us. An entire universe that has scarcely been explored lies before the scientist. There is the sky sprinkled with celestial bodies moving about in the darkness of infinite space, the sea with its mysterious depths, the earth guarding within its innermost recesses the history of life, including man's predecessors; and finally, the human organism or masterpiece of creation. Each cell presents us with the unknown, and each heartbeat inspires profound meditation within us.

Carried away by enthusiasm, perhaps I am slipping into hyperbole. But I am convinced that true originality is found in science, and that the fortunate discoverer of an important fact is the only one who can flatter himself with having trodden on completely virgin territory—and with having forged a thought that never before passed through the human mind. And let me stress that this conquest of ideas is not subject to fluctuations of opinion, to the silence of envy, or to the caprices of fashion that today repudiate and detest what yesterday was praised as sublime. James's thought applies especially to the fortunate inquirer of nature: man's ideal lies in collaborating with God.

It is certainly true that the scientist's fame is not as great as the playwright or artist's glamour and popularity. People live in a world of sentiment, and it is asking too much of them to provide warmth and support for the heroes of reason. Nevertheless, scholars do have their public. It consists of the intellectual aristocracy that dwells in every country, speaks every language, and reaches the most distant of future generations. These admirable men of science do not, of course, applaud or forget themselves in torrents of emotion. Instead, they study with affection, judge with moderation, and end by rendering full and final justice—ignoring for the moment fleeting attacks of envy. As for reputation, the supreme accomplishment is to merit the approval of those rare and superior minds that humanity produces from time to time. This allows one to understand the nobly high-minded sentiment behind a comment of the mathematician and philosopher Fontenelle to someone after presenting his treatise on the *Geometrie de l'infini:* "This is a work that only four or five people in France will be able to read." The well-known statement Kepler made in closing his work *Harmonices mundi* is also noble and packed with feeling. Radiating joy and trembling with emotion over the discovery of the last of his memorable laws, Kepler wrote: "The die is cast, and with this I finish my book, caring little whether it is read today or by posterity. Some day there will be readers. After all, did God not wait six thousand years to find in me a beholder and interpreter of His works?"

Patriotism

Patriotism merits special attention as one of the emotions that should inspire the man of science. For him, it has an entirely positive connotation: he is eager to enhance the prestige of his country, but without destroying the reputation of his contemporaries.

It has been said that science has no country, and this is true. But as Pasteur once replied on a solemn occasion, "Scientific scholars do have a country." Nature's conqueror belongs not only to humanity but to a race that shows pride in his talents, to a nation that is honored by his triumphs, and to a region that considers him the choice fruit of its soil.

Because science and philosophy represent the highest branches of mental activity, and serve as the dynamometers for human spiritual energy, it is clear why civilized nations exhibit noble pride in their philosophers, mathematicians, physicists, naturalists, and inventors—in short, all of those who know how to exalt their country's honored name.

It must be confessed that we Spaniards have the greatest need for cultivating this emotion because of the many centuries of disdain that we have given anything dealing with scientific investigation and its fruitful applications to life (for reasons that are now irrelevant). All of us who wish to keep the fiber of patriotism supple—after repeated injuries from the darts of foreign indifference—have a fundamental obligation to defend the prestige of the Spanish people. We must prove to foreigners that those who centuries ago knew how to immortalize their names and compete with the eminent nations in deeds of war and in the perils of exploration and geographic discovery, just as in the peaceful undertakings of art, literature, and history, can also struggle with equal tenacity and energy in the investigation of nature, collaborating harmoniously with the best-informed nations in the great work of civilization and progress.

Some thinkers, including Tolstoy, are inspired by humanitarian feelings that are as much at variance with reality as they are inopportune in these times of cruel international rivalries. They have stated that patriotism is an egocentric sentiment that inspires endless warfare and is destined to vanish—replaced by the more noble and altruistic ideal of universal brotherhood.

One must recognize that when patriotic fervor extends to chauvinism, it creates and maintains exceptionally dangerous rivalry and hatred among nations. However, within prudent limits, and tempered by the justice and respect due the science and power of foreign nations, it must be admitted that patriotism encourages the highest standards of international competition—and advances the causes of progress and humanity. International scientific congresses are very profitable when regarded in this way because many scholars distrust one another at the beginning, perhaps because of international rivalries or perhaps because of the noble and laudable form of envy championed by Cervantes. However, coming into contact allows them to get to know one another and develop a cordial respect. Furthermore, currents of sympathy and justice beginning at the highest levels quickly filter down to the very center of the social masses, progressively easing political tensions between rival peoples.[1]

At any rate, loyalty to country will always exert its dynamic force and provide a major stimulus to scientific and industrial competition, however cosmopolitan the world may become. The psychological roots of patriotism are buried too deep. They cannot be destroyed by the attacks of international socialism and the ponderous studies of philosophical humanism. Passions of this kind are not discussed, they are utilized—because they generate invaluable stores of energy and sublime acts of heroism. The mission of governments and educational institutions is to channel or tame this admirable force, applying it to useful and redeeming enterprises and steering it away from the attacks and turmoil of murderous dissension.

P.J. Thomas has made the wise observation in his *Education of the Emotions* that "the idea of country, like the idea of family, is necessary, as are the feelings involved in both. They work like stimulants to progress, and they guarantee our own dignity. One struggles for the glory of his country just as one struggles for the honor of his name . . . The nation, it has been said, is an indestructible element in the harmony of the world, having equal title with the province, the family, and the individual . . . Humankind must retain diversity to keep itself strong and promote continuous regenerative activity."

Even in the improbable event that a United States of Europe or the World forms, man will always love above all his worldly, moral, and material environment—that is, his church spires, his own community, and his race—and will devote lukewarm affection bordering on indifference to more distant influences. It has often been said with good reason that man's attachment to and affection for material things is inversely proportional to the distance in time and space between them. "Time" is important because our country is not only home and native soil, it is also past and future—our remote ancestors and our future descendants.

Bayle cogently pointed out that "Our resolve to work does not come from general ideas formed in the mind, it comes from emotions formed in the heart." And among these emotions, none holds in its annals more glorious deeds than love of country. It matters very little whether such feelings are just or unjust, or whether they reveal a primitive and barbarous side of humanity. They are intellectual stimulants that should be judged strictly by their effects, or, as is fashionable these days, pragmatically.

Taste for Scientific Originality

The drive to carry out great undertakings may certainly come from the inducements of patriotism and a lofty desire for reputation. Nevertheless, our novice runs the risk of failure without additional traits: a strong inclination toward originality, a taste for research, and a desire to experience the incomparable gratification associated with the act of discovery itself.

The important roles of scrutinizing mysteries and investigating new observations in mobilizing action have been discussed often. Eucken has written memorably about this, keenly noting that "action *personalizes* us; it creates the pleasant illusion of being sovereign creators, and with a feeling of unshackled freedom, the joy of unlimited power."

Aside from increased feelings of self-esteem and the approval of our own conscience, the conquest of a new truth is undoubtedly the greatest adventure to which man can aspire. The cajoleries of vanity, the effusions of instinct, and the caresses of fortune pale before the supreme pleasure of experiencing how the wings of the spirit emerge and develop, and how when working harmoniously we overcome difficulty to dominate and subdue elusive nature.

The man of science can defy even injustice when fortified by this hedonic sentiment. His enthusiasm will not allow him to be injured by the deliberate silence of his rivals who, as Goethe pointed out, often attempt to ignore what they would like to remain unknown. Nor will it allow him to be injured by the fact that contemporary society does not understand him, or that he is forgotten by the official institutions. The regard that the world has for power, aristocracy, and money is not a prime aspiration for him because he feels within himself a nobility greater than that granted capriciously by blind fortune or the good humor of a ruler. He is as proud of this nobility as he is of his own work. Through it, he is a minister of progress, a priest of truth, and a confidant of the Creator. He is devoted entirely to understanding something of that mysterious language that God has written in nature. He alone is allowed to penetrate the marvelous works of Creation. He renders to the Absolute the most pleasing and acceptable homage—studying His prodigious handiwork so as to know, admire, and revere Him through it. Stooping to the trifles of human egotism, everyone would agree that we are esteemed and respected only by those who study and try to understand us.

As we have already noted, the joyful emotion associated with the act of discovery is so great that it is easy to understand the sublime madness of Archimedes. Historians tell us that he was so beside himself after solving a profound problem that he burst forth almost naked from his house exclaiming the famous *Eureka*, "I have found it!"

Recall the happiness and emotion displayed by Newton when his genius-inspired conjecture about universal attraction was confirmed by calculation, and by Picard's measurements of a terrestrial meridian. No matter how modest, every investigator has probably at one time or another felt something of the immeasurable satisfaction that Columbus must have experienced when he heard Rodrigo de Triana shout "Land! Land!"

This indescribable pleasure—which pales the rest of life's joys—is abundant compensation for the investigator who endures the painful and persevering analytical work that precedes the appearance of the new truth, like the pain of childbirth. It is true to say that nothing for the scientific scholar is comparable to the things that he has discovered. Indeed, it would be difficult to find an investigator willing to exchange the paternity of a scientific conquest for all the gold on earth. And if there are some who look to science as a way of acquiring gold instead of applause from the learned, and the personal satisfaction associated with the very act of discovery, they have chosen the wrong profession! They should wholeheartedly dedicate themselves to the exercise of industry or commerce instead.[2]

In fact, beyond the stimulation of variety and interest, the supreme joy of the intellect lies in seeing the divine harmony of the universe, and in knowing the truth—as beautiful and virginal as the flower opening its calyx to the caresses of the early morning sun. As Poincaré says in his beautiful book, *La Science et la methode,* "Intellectual beauty is sufficient unto itself, and only for it rather than for the future good of humanity does the scholar condemn himself to arduous and painful labors."

Notes

1. This frank optimism is now greatly undermined by the hideous international war that began in 1914. When this was written in 1893, everything pointed to the belief that the era of great European wars had passed. Railways, the telegraph, newspapers, congresses, international conferences, the spread of languages—all seemed to be instruments destined sooner or later to help realize the noble aspiration of uniting the European nations, or of at least bringing them closer together.

It was consoling to see the hands of philosophers, scholars, and workers joined cordially regardless of political frontiers. Unfortunately, military governments and insatiable profiteers acted in the opposite way and, thanks to intense inoculation begun in the schools, relentlessly smothered the seed of love with the venom of hatred. The task, which is perhaps a chimera, of permanently reconciling the states of Europe and bringing an end to their atavistic covetousness and unruly territorial ambitions, will fall anew to the twenty-first century.

This was written in 1916. Today the peace has been signed, Europe is in ruins, the naive Wilsonian concept of the League of Nations has failed, and the hatred of conquered people dreams of retaliation. We look with bitter skepticism at all juridical plans for lasting peace. It is sad to admit, but all nations become ferociously imperialistic as soon as possible, whether they be monarchies or republics. So much for the weak and unpatriotic!

2. Things are now somewhat different. The type of inventor whose efforts are driven by an eagerness for monetary gain is now common in Germany and, generally, in the other advanced nations. The struggle for patents, and the fever of industrial competition, have disturbed the august calm in the temple of Minerva. Is this good or bad?

4 What Newcomers to Biological Research Should Know

General education. The need for specialization. Foreign languages. How monographs should be read. The absolute necessity of seeking inspiration in nature. Mastery of technique. In search of original data

General Education

We needn't dwell on the fact that our novice must acquire a thorough knowledge of the science intended to be explored. It is essential that this knowledge come from descriptions in books and monographs, as well as from the direct study of nature itself. However, it is equally important that he acquire a general knowledge of all those branches of science that are directly or indirectly related to the one of choice because they contain guiding principles or general methods of attack. For example, the biologist does not limit his studies to anatomy and physiology, but also grasps the fundamentals of psychology, physics, and chemistry.

The reason for this general education is obvious. The discovery of a fact or the significance of a biological phenomenon usually rests on the application of principles derived from physics or chemistry. As Laplace has pointed out, to discover is to bring together two ideas that were previously unlinked. And it is important to note that this fruitful association typically occurs between data from one of the complex sciences (biology, sociology, chemistry, and so on) and a principle culled from a general science. In other words, the general or abstract sciences (according to the classification of Comte and of Bain) often explain the phenomena of the complex and concrete sciences, leading one to conclude that a well-established hierarchical classification of human knowledge constitutes a veritable genealogical tree.

Discovery is often a matter of simply fitting a piece of data to a law, or wrapping it in a broader theoretical framework, or, finally, classifying it. Thus, it may be concluded that to discover is to name something correctly, something that had been christened incorrectly or conditionally before. This leads to the conclusion that when science has been completed, each phenomenon will have its correct name, after its relationship to general laws has been firmly established. When viewed in this way, the well-known saying of Mach acquires real meaning: "A well-chosen word can save an enormous amount of thought," because *to name* is to classify, to establish ideal affiliations—analogous relationships—between little-known phenomena, and to identify the general idea or principle wherein they lie latent, like the tree within its seed.

More than anything, the study of philosophy offers good preparation and excellent mental gymnastics for the laboratory worker. Do not forget that many renowned investigators have come to science from the field of philosophy. It is needless to point out that the investigator will be concerned less with doctrine and philosophical creed—which unfortunately change every fifteen or twenty years—than with the criteria of truth and the standards of critical judgment. Exercise of the latter allows one to acquire flexibility and wisdom, as one learns to question the apparent certainty of the best-established scientific systems. This is how one's own imagination is properly bridled. The investigator's motto will always be Cicero's phrase: *Dubitando ad veritatem pervenimus.*

As far as the microscopic anatomy of plants and animals is concerned, most of the core data forming this science are derived from interactions between the chemical properties of certain reagents and the structural elements of cells and tissues. In bacteriology, neurology, and so on, we owe most of what we know today to the fortunate application of the staining agents developed by modern chemistry, and the same applies to general biology. Simply recall Loeb's interesting work on artificial parthenogenesis, or the work of Harrison, Carrel, Lambert, and others on artificial cell cultures from animal tissues. Novel experiments such as these rely fundamentally on chemical and physical changes taking place in the cellular environment.

This intimate union of the sciences has been appreciated by many. It was especially clear to Letamendi, who defined scientific specialties as "the application of science as a whole to a particular branch of knowledge."

If a supreme intelligence knew all the mysterious explanations linking all phenomena in the universe, there would be *one single science* instead of many different sciences. The frontiers that appear to separate fields of learning, the formal scaffolding of our classification scheme, the artificial division of things to please our intellects—which can only view reality in stages and by facets—would disappear completely in the eyes of such an individual. Total science would appear as a giant tree, whose branches represent the individual sciences and whose trunk represents the principle or principles upon which they are founded. The specialist works like a caterpiller perched on a leaf, cherishing the illusion that his little world flutters isolated in space. Endowed with a philosophic sense, the generalist sees—however imperfectly—the stem that is common to many branches. But only the genius alluded to above would enjoy the good fortune and power to see the entire tree, s*cience,* unitary despite its many specializations.

The Need for Specialization

It is wise, however, not to emphasize the unifying principle just discussed. It is too easy to run aground on the shoal of encyclopedic learning, where minds incapable of orderliness—who are restless, undisciplined, and unable to concentrate attention on a single idea for any length of time—tend to stop. *Rotating inclinations,* as a highly original physician-writer has called them, may create great writers, delightful conversationalists, and illustrious orators, but rarely scientific discoverers.

The well-known proverb, "Knowledge does not occupy space," is a grave mistake. Fortunately, this is of little practical consequence because even those who believe it must confess that learning many things at the very least takes time. Only an excessively flattering estimate of one's talents can explain the encyclopedic mania. The intent to master a number of sciences is a chimerical aspiration. Just consider the indefatigable men of real genius who resign themselves to a profound knowledge of one branch of knowledge—and often to one concrete theme within a given science—only to harvest a small number of facts.

In short, do not get carried away by illusion. If a lifetime is needed to learn something about all of the human arts, it is barely sufficient to master completely, down to the last detail, any one or two of them.

Modern encyclopedists such as Herbert Spencer, Mach, and Wundt are actually specialists in the philosophy of the sciences and arts, as were Leibnitz and Descartes in their own time. However, the latter were able to dominate larger territory, and make discoveries in two or three sciences, because less was known during their lifetimes.

Multifaceted investigators have disappeared, perhaps forever. It is important to realize that today, in physics as in mathematics, in chemistry as in biology, discoveries are made under the astute direction of specialists. However, they do not focus exclusively on a narrow topic; instead, they follow attentively the latest developments in related sciences, without losing sight of their specialty. In addition to being good policy, this division of work is an indispensable necessity. We are forced to adopt it because of the extraordinary amount of time required by the testing and mastery of new techniques reported almost daily, by the growing volume of the literature, and by the many scholars working simultaneously on virtually every topic.

It seems useful to conclude with a frequent comparison that is expressed in two maxims of everyday philosophy: "A long harvest, little corn," and "Knowledge occupies no space." The inquisitive mind is like a sword used in battle. If it has one sharp edge, we have a cutting weapon; with two edges it will still cut, though less efficiently; but if three or four edges are arranged simultaneously, the effectiveness of each diminishes progressively, until the sword is reduced to a dull bludgeon. Strictly speaking, a bayonet still cuts, although a great deal of energy is required, whereas a well-sharpened dagger is dangerous even in the hands of a child.

Like unmolded steel, our mind represents a potential sword. The forging and polishing of study transform it into the tempered and keen scalpel of science. Let us have a cutting edge on only one side, or on two at the most, if we want to conserve its analytical powers and penetrate to the heart of problems. Leave to the scatterbrained encyclopedists the privilege of transforming their minds into dull weapons.

Foreign Languages

Obviously, the investigator's library should contain the important books and journals related to his specialty that are published in the most advanced countries. The German journals will be consulted regularly because it must be admitted that Germany alone produces more new data than all the other nations combined when it comes to biology.[1]

He who desires an end desires the means. Because a knowledge of the German language is essential to keep abreast of the latest scientific news, let us study it seriously, at least to the point of being able to translate it adequately. Abandon the superstitious terror that the complicated twists and turns of the Northern languages inspires in us Spaniards. A knowledge of German is so essential that today there is probably not a single Italian, English, French, Russian, or Swedish investigator who is unable to read monographs published in it. And because the German work comes from a nation that may be viewed as the center of scientific production, it has the priceless advantage of containing extensive and timely historical and bibliographic information.[2]

In order of importance, English and French follow German. We shall not discuss Italian because any Spaniard with an average education can translate it, even without the help of a dictionary. However, it is important to recall that in some areas of scientific work, Italy marches at the head of the procession.

At the present time scientific work is published in more than six languages. Although reverting to Latin or using Esperanto as a universal scientific language might appear useful, scholars have responded instead by using even more languages for presenting their scientific work in print. It must be admitted that from the standpoint of practicality, Volapük or Esperanto simply became one more language to be learned.[3] This reality should have been predicted because it is inevitable in this day and age, with the essentially democratic and popularizing influences of modern knowledge, and the practical views of authors and editors whose moral and material interests impel them to spread among the general public the scientific victories that long ago were the exclusive property of the academies and a tiny group of lecture hall celebrities.

It is clearly not necessary, however, for the investigator to speak and write all of the European languages. It is enough for the Spaniard to translate the following four: French, English, Italian, and German. It is appropriate to call them the *languages of learning,* and virtually all scientific work is published in one of them. Naturally, Spanish does not figure among the languages of learning. Therefore, if our investigators want their research to be known and appreciated by the specialists, they have no choice but to write and speak one of these four European languages.

How Monographs Should Be Read

When monographs on the specialty one has chosen to investigate are read, particular attention should be focused on two important things: research methods used by the author in his work, and problems that remain unsolved. What might be called popularizing books merit less attention and confidence, unless they are comprehensive reviews of the field of interest, or contain useful general concepts that may be applied in the laboratory. In general, it may be assumed that books reflect historical eras in science. Because they take so long to write and edit, especially when authors are determined to simplify the material so that the public can readily understand it, books are either not written about contemporary issues, or they are very lightly sketched. The same holds true for methodological details and paths of investigation.

Monographs by outstanding authors who have contributed most to a field should be submitted to thorough and critical study. Among other qualities, original talent has the great virtue of stimulating thought. A feature of every good book is that it allows the reader not only to extract the ideas deliberately presented by the author but also to formulate completely new ideas (different for each reader) that result from a conflict between our own fund of ideas and views expressed in the text. Clearly, the good treatise is an effective reagent for our cerebral energies, in addition to being an excellent source of scientific lore.

Human brains, like desert palms, pollinate themselves at a distance. However, for the union of two minds to occur and generate fruitful results through a book, the reader must become fully absorbed in what a master has written, must penetrate fully its meaning, and finally must develop an affection for the author. In science as in life, fruit always comes after the realization of love. So many beginners fall into the trap of considering old or even ancient discoveries as the fruit of their own labor simply because they relied on secondary sources instead of consulting original reports!

Our new and inexperienced man of science must flee from abstracts and syllabuses as if from the plague. The syllabus is good for teaching purposes, but is abominable for guiding the investigator. He who abstracts a book does so with his own purposes in mind. He often displays his own judgments and doctrines instead of the author's. He takes from the latter what pleases him or what he understands and digests easily. And he makes important that which should be secondary, and vice versa. In the name of clarifying and popularizing someone else's work, he who does the abridging ends up substituting his own personality for that of the author, whose intellectual character—which is so interesting and educational for the reader—remains in the shadows.

Thus, the investigator has a strict obligation to read an author's original work if he wishes to avoid disagreeable surprises—unless the abstract is by the author himself. Here at least, we may find original and guiding ideas that can be used to real advantage in analytical work, despite their brevity.

At this point, an interesting question emerges: should beginners review the literature before starting experimental work? Permeated, if not saturated, by whatever has been written on the subject, mightn't we run the risk of being influenced, and of losing the invaluable gift of independent judgment? Won't our aspirations of finding something completely original be fatally injured by all the detailed information we have surrendered ourselves to, leaving the impression that there is nothing left to discover?

Each person must solve this problem in his own way. Nevertheless, there seems to be general agreement that no inquiry should be started without having all the relevant literature at hand. This approach avoids the painful disillusionment that comes with finding that we have squandered our time rediscovering something already known, thus neglecting the profound study of genuinely unknown aspects of a problem.

In my view, the wisest course is to complete a thorough review of the literature routinely before launching an analytical project. But when this is not feasible due to insuperable difficulties (which unfortunately occur in Spain, where the universities lack recent foreign books, and the academies do not have the resources to subscribe to the most important scientific journals), we should not desert the laboratory because one or another monograph is unavailable. If we work hard and long with the best methods available, we shall find something that has escaped the attention of previous workers. Not having been influenced by them, we shall have traveled different routes and considered the subject from different points of view. In any event, it is worth a thousand times more to risk duplicating discoveries than to give up all attempts at experimental investigation. This is true because when a beginner's results turn out to be similar to those published a short time earlier, he should not be discouraged—instead, he should gain confidence in his own worth, and gather encouragement for future undertakings. In the end he will produce original scientific work, providing his financial resources match his good intentions.

The Absolute Necessity of Seeking Inspiration in Nature

We may learn a great deal from books, but we learn much more from the contemplation of nature—the reason and occasion for all books. The direct examination of phenomena has an indescribably disturbing and leavening effect on our mental inertia—a certain exciting and revitalizing quality altogether absent, or barely perceptible, in even the most faithful copies and descriptions of reality.

All of us have probably observed that when we attempt to verify a fact presented by a writer, unexpected results invariably emerge, suggesting ideas and plans of action not aroused by the mere act of reading. In our view, this is due to an inability of the human word to paint exactly and faithfully. In any branch of knowledge one may wish to mention, reality presents a surface of highly varied and complex sensations. Symbolic expression always arises through abstraction and simplification, and can only reflect a small part of reality.

No matter how objective and simple it may appear, all description relies on personal interpretation—the author's own point of view. It is well-known that man projects his personality onto everything, and that when he believes he is photographing the outside world he is often observing and depicting himself.

From another perspective, observation provides the empirical data used to form our conclusions, and also arouses certain emotions for which there are simply no substitutes—enthusiasm, surprise, and pleasure, which are compelling forces behind constructive imagination. Emotion kindles the spark that ignites cerebral machinery, whose glow is required for the shaping of intuition and reasonable hypotheses.

As an example of the direct thought-provoking effects that nature has on the observer, it seems appropriate to relate the impressions I felt on observing the phenomenon of the circulation of the blood for the first time.

I was in my junior year of medical studies and had learned about the details of this phenomenon from various books, although my interest had not been aroused particularly, and I had given it little thought. However, one of my friends, Mr. Borao (a physiology assistant), was kind enough to demonstrate the circulation in the frog's mesentery to me.

During the sublime spectacle, I felt as though I were witnessing a revelation. Enraptured and tremendously moved on seeing the red and white blood cells move about like pebbles caught up in the force of a torrent; on seeing how the elastic properties of red corpuscles allowed them suddenly to regain their shape like a spring after laboriously passing through the finest capillaries; on observing that the slightest obstruction in the stream converted potential spaces between endothelial cells into actual spaces providing the opportunity for minor hemorrhage and edema; and finally, on noticing how the cardiac beat weakened by curare slowly propelled the obstructing red corpuscles, it seemed as though a veil had been lifted from my soul, and my beliefs in I know not what mysterious forces I had attributed the phenomena of life receded and vanished. In my enthusiasm I exclaimed the following, not knowing that many others, including Descartes especially, had done so centuries before: "Life seems to be pure mechanism. Living bodies are hydraulic machines that are so perfect they can repair the damage caused by the force of the torrent moving them, and even produce other similar hydraulic machines through the mechanism of reproduction." I am absolutely convinced that the vivid impression caused by this direct observation of life's internal machinery was one of the deciding factors in my inclination to biological research.[5]

Mastery of Technique

Once a topic has been chosen for study, and the investigator has examined all of the literature relevant to the problem of interest, it is time to confirm the latest published data with the most appropriate analytical methods available. Very often during this attempt at proof, questionable points, untenable hypotheses, and gaps in observation will be recognized; and now and then the young investigator may glimpse the road he will be privileged to travel along in search of knowledge about the problem.

Mastery of technique is so important that without fear of contradiction it may be stated that great discoveries are in the hands of the finest and most knowledgeable experts on one or more of the analytical methods. Through intense application, the masters have learned all of the secrets that the technique may have to offer.

In support of this contention, I would simply ask you to recall that of the hundreds of histologists, embryologists, and anatomists working in Europe and America, the most important scientific conquests have been won by only a dozen men who became known for their invention or improvement of a research method, or for their having absolutely mastered one or more of them.

The latest research techniques can be given preference, but first priority must go to the most difficult because they are the least exploited. Time wasted on experiments that don't work does not matter. If the method has very high resolution, the desired results will have real importance, and will repay our eagerness and zeal quite handsomely. Moreover, difficult techniques provide us the inestimable advantage of proceeding almost alone, finding very few imitators and competitors along the way.

In Search of Original Data

We now come to a difficult problem: the great frustration of the beginner who knows from the history of scientific research that once the first discovery is made, others related to it will certainly follow.

A new discovery is often the fruit of patient and stubborn observation—the result of having spent more time, been more consistent, and used better methods than our predecessors. As we have already pointed out, scrupulous and repeated consideration of the same data eventually yields a supersensitive, refined, analytical perception of whatever is relevant to the chosen problem. How often we find entirely new things in preparations, where our unsuspecting pupils saw nothing! This is due to the quick judgment that results from experience. And how many things probably escaped our attention when we were still inexperienced in microscopic technique and each preparation appeared like a sphinx defying understanding!

In addition to how remarkably our differentiating powers increase through repeated experimentation and observation, the resolute study of a problem almost always suggests improvements in methodology, after one determines the conditions that produced some unfortunate result and thus the factors that yield maximum technical efficiency.

Diligence is often rewarded with discovery. It is simply a matter of applying a recent technique to a problem that has lain dormant for some time. These tactics have generated tremendous and easy progress in bacteriology and in comparative anatomy and histology.

Because the great pioneers of science typically have created new methods, it would be ideal if rules for their discovery could be formulated. Unfortunately, almost all of the analytical methods in biology have been found as a result of chance.

It is true to say that, in general, methods are useful applications to one branch of knowledge of principles belonging to another. However, the applications are often developed by trial and error, or at most are inspired by vague analogies.

As we have already noted, in areas such as bacteriology, histology, and histochemistry, methods are based on the selective effects of dying agents or of reagents created by modern chemistry. However, there was no rationale—unless the intention was to produce a useful result by chance—for Gerlach to stain nuclei with carmine; for Max Schultze to use osmic acid on nervous tissue; for Hannover to fix tissues with chromic acid and dichromates; for Koch, Ehrlich, and others to stain bacteria with the aniline dyes; and so on.

If we knew the entire chemical composition of living cells, results due to the application of a particular staining reagent could be deduced simply from biochemical principles. However, because we are so far from this position, those aspiring to discover new biological methods are forced to submit live tissues to the same blind tests resorted to by chemists for centuries in the hope of now and then finding some unforseen combination of reactions or mixtures of elements.

Thus, it is necessary to trust at least partly in chance, which can be encouraged by repeated series of trials that must be guided by intuition and as deep and accurate a knowledge as possible of the latest reagents and techniques emerging from chemistry and industry.

This brings me to the point of discussing the role of chance in the realm of scientific investigation. There is no doubt that accident is a major component of empirical work, and we must not overlook the fact that science owes brilliant achievements to it. However, as Duclaux has graphically pointed out, chance smiles not on those who want it, but rather on those who deserve it. It is important to recognize that only the great observers benefit from chance because only they know how to pursue it with the necessary strength and perseverance. And when an unexpected revelation appears, only they are in a position to realize its great importance and scope.

In science as in the lottery, luck favors he who wagers the most—that is, by another analogy, the one who is tilling constantly the ground in his garden. If Pasteur discovered bacterial vaccines by accident, he was assisted by genius. He envisioned all of the benefits that might be derived from a casual observation, the reduced virulence of a bacterial culture exposed to air (probably reduced by the action of oxygen).

The history of science is full of similar tales. Scheele happened upon chlorine while trying to isolate manganese; Claude Bernard planned experiments to characterize the destructive agent in sugar but instead discovered the glycogenic function of the liver; and so on. To end this section, we shall consider two recent examples of almost miraculous good luck in the stupendous discoveries of Roentgen, Becquerel, and the Curies.

It is well known that the discovery of x-rays by Professor Roentgen was the result of simple chance. In his Würzburg laboratory, this learned man repeated the experiments of Lenard on the unique properties of *cathode rays*. The emitted radiation was projected in the usual way onto a screen made fluorescent with barium platinocyanide. Roentgen was interested in determining how long the fluorescence lasted, and one day it occurred to him to darken the laboratory by covering the Crookes tube (the well-known apparatus that generates cathode rays) with a cardboard box. When the transformer was turned on, Roentgen looked at the screen and to his amazement saw that it was brightly illuminated anyway. He then substituted a piece of wood, and then a book, and he observed that part of the radiation—the new rays—went through these opaque objects readily. Finally, in a moment of feverish impatience, he accidentally placed his hand between the Crookes tube and the fluoroscopic screen. Overcome with intense emotion—perhaps even terror—he observed an astonishing spectacle: on the surface of the fluorescent screen, the bones of his hand were faithfully depicted in black, as if there were no surrounding tissues. The wonderful x-rays had been discovered, and with them *fluoroscopy*. *X-ray photography* and the many valuable surgical and industrial applications known to all soon followed.

The second story is just as elegant, and involves the accidental discovery of *radioactivity*, which we owe to the renowned French physicist, Henri Becquerel.

The unappreciated Poincaré had already asked whether it might not turn out in the long run that x-ray production is a property of all fluorescent elements. Wanting to confirm this suggestion, and well prepared for this line of investigation, M. Becquerel planned to examine *uranium sulfate*, a typical fluorescent compound. But the cloudy days of February came and went, and the sun failed to appear. Hoping the celestial monarch might dissipate the thick Parisian mists, the physicist prepared his experiment with a great deal of anticipation. He placed various crystals of uranium sulfate on a sensitized plate that was wrapped in black paper, and he also placed a copper cross on the plate. Impatience devoured and goaded him until one day it occurred to him by chance to remove the plate from its protective wrapping and develop it. To his great surprise, and contrary to his expectations (the uranium salts had remained in the dark), a vivid pattern was observed on the plate: The crystals of uranium salts were represented in black and the metallic cross was represented in white. Without intending to, he had discovered the emission of radioactivity by matter, one of the most wonderful conquests of modern science.

But the most extraordinary and outstanding part of this story is that M. Becquerel was able to make such a great discovery (winning him a Nobel prize) on the basis of a false hypothesis—that there is an etiologic relationship between the emission of x-rays and fluorescence. By sheer coincidence, of all known fluorescent elements, uranium is the only one that is radioactive! The effect was obviously theatrical, as if prepared by an ironic genius determined to prod science along in spite of how inaccurate its concepts may be.

It is obvious that while many scholars have discovered things they were not actively looking for, they nevertheless searched with admirable tenacity and were worthy of their success. With rare insight they succeeded in finding unexpectedly the great developments that lie hidden within the timid and fragmentary revelations of the great unknown. In the long run, fortunate accidents almost always have a way of rewarding perseverance.

To solicit the aid of luck is like stirring muddy water to bring objects submerged at the bottom to the top where they can be seen. Every observer would do well to tempt their good luck. Nevertheless, we should not depend on it too much—most of the time is it better to concentrate on systematic work. He who masters technique and keeps abreast of problems that can be solved almost always comes away with a more or less important discovery without doing a lot of unproductive experiments.

Once the first new data are obtained—and especially if they generate new currents in the scientific atmosphere—our task will be as smooth as it is brilliant because it reduces to working out the consequences of the new data for the various spheres of science. This leads to the conclusion that the first discovery is the one that counts; the rest are usually corollaries of the original. A well-known doctrine espoused by philosophers such as Taine, and by scientists such as Tyndall, is that each problem solved stimulates an infinite number of new questions, and that today's discovery contains the seed of tomorrow's. The peak of truth that took so much effort to scale appears to be an imposing mountain when gazed at from the valley. However, it proves to be just another mountain within a formidable succession of ranges, seen through the mist and attracting us with insatiable curiosity. We should satisfy our eagerness to climb, and take advantage of the peace one experiences in contemplating new horizons. And from the recently conquered peak, also think about the path that leads to even higher ranges.

But as we have said before, it is very rare indeed to have the good fortune of starting out with a promising study that actually produces an important discovery, and no wise investigator counts very much on doing so. Therefore, we must not hesitate when beginning our work to follow up someone else's discovery. This is a useful task, and useful results will follow. Original data produced by others often foment revolution in the scientific atmosphere. They raise doubts about what were considered established principles; they change the equilibrium in those vague regions of conjecture that shape the transition from what is unknown to what is known; and they establish a new set of problems that the discoverer himself lacks the time to pursue.

Furthermore, the original investigator almost always leaves his discussion incomplete. He is still influenced by tradition and fails to break openly with the views of the past. Perhaps wary of arousing too much opposition in the scientific world, and anxious for approval and applause, his theory is presented as a compromise between the oldest and the newest doctrines. Thus, a less meticulous newcomer could well advance the work of the originator with little effort, and obtain the most valuable theoretical and practical results. The many problems raised by a new scientific discovery are fertile ground for the young investigator. His analytical weapons finely honed, he will respond without arrogance or great expectations—but must not count on arriving alone. He will find a host of rivals attempting to surpass him, and can excel only by virtue of hard work, clear-sightedness, and perseverance.

Finally, when we discover ourselves surrounded by a number of equally promising and fertile problems to work on, choose the one whose methodology we understand clearly, and the one we have a decided liking for. This is the good advice Darwin used to give his students when they asked for a problem to work on. The rationale for this approach is that our intellect redoubles its efforts when perceiving the reward of pleasure or utility in the distance.

As we have already pointed out on a number of occasions, the explorer of nature must view research as the best of all possible sports, whose every facet—from the execution of technique to the elaboration of theory—is a never-ending source of indescribable satisfaction. You should abandon science if you don't feel growing enthusiasm and a growing sense of power when working with a difficult problem—if your soul isn't flooded with the emotion of anticipated pleasure when approaching the long-awaited and solemn moment of the *fiat lux*. Nature grants not her favors to those with a cold heart—which is usually an unmistakable sign of impotence.

Notes

1. Because of growing rivalries, centers producing biological research are springing up in a number of places. Italy, France, England, and especially

the United States, are now competing with, and in many cases surpassing, the work of German universities, which for decades was incomparable. (Author's footnote, 1923).

2. When the Spanish attend scientific congresses, they deplore the fact that their language is eclipsed by German, French, or English. Before registering complaints that automatically evoke smiles from the learned, these inopportune patriots would be well advised to think seriously about the following three undeniable observations:

a. Qualitatively and quantitatively, our scientific production is much less than that of the four nations enjoying the privilege of using their native tongues at congresses.

b. As a result, Spanish is unknown to the vast majority of scholars. If a quixotic patriotism inspires us to insist on using it at our international congresses, we should expect a mass exodus of listeners.

c. Finally, nations including Sweden, Holland, Denmark, Hungary, Russia, and Japan—whose scientific production far exceeds that of Spain—have never been guilty of boldly imposing their respective languages on such literary contests. Their scholars are far too clear-sighted not to recognize that while the task of mastering the four languages under consideration is difficult, learning one or two more languages is not an intolerable form of torture.

3. If international jealousies and suspicions could be avoided, it would be much more simple and practical to agree on the use of a living language—French, for example—for scientific communication. It would be interesting to ask Esperanto enthusiasts whether they plan to abandon the use of French while traveling in France. (As one might have predicted, the brand new Volapük has already [1920] been forgotten completely. I predict the same for Esperanto.)

4. Thanks to laudable initiative, the German language has been given special attention in the curriculum of our institute. Unfortunately, however, this has as yet done little good for our scholars. This is due as much to a lack of time for the subject as to shortsighted teaching methods. When there is not enough time to master a difficult language, it would make sense not to attempt teaching *all aspects of German.* Instead, it would seem logical to concentrate on scientific German—on the relatively limited set of grammatical rules and small number of voices needed to translate scientific monographs. This can be achieved with six to eight months of diligent study. We suggest to those with an interest in biological work to subscribe to a German journal in his specialty, for example, any *Zentralblatt.* While laborious at first, the reading will come easier. The pleasure of obtaining

some benefit from the very outset will progressively add to the enjoyment of his research.

5. Today I do not subscribe unreservedly to this mechanistic concept, nor do I adhere strictly to the physicochemical interpretation of life. The origin and morphology of cells and organs, heredity, evolution, and so on include phenomena that depend on incomprehensible absolute causes, notwithstanding the vaunted promise of Darwinism and the postulates of Loeb's school of biochemistry.

5   Diseases of the Will

Contemplators. Bibliophiles and polyglots. Megalomaniacs. Instrument addicts. Misfits. Theorists

We have all seen teachers who are wonderfully talented and full of energy and initiative—with ample facilities at their disposal—who never produce any original work and almost never write anything. Their students and admirers wait anxiously for the masterpiece worthy of the lofty opinion they have formed of the teacher. But the great work is never written, and the teacher remains silent.

Let us not be deceived by optimism and good intentions. Despite their exceptional merit, and the zeal and energy they display in the classroom, such teachers suffer from a disease of the will—although psychologists may not see it this way. Their sluggishness and neglect may not justify a diagnosis of abulia or loss of will power, but their students and friends may nevertheless consider them abnormal and suggest some adequate form of spiritual therapy, with all due respect to their fine intellectual abilities.

These illustrious failures may be classified in the following way: t*he dilettantes or contemplators; the erudite or bibliophiles; the instrument addicts; the megalomaniacs; the misfits;* and *the theory builders.*

Contemplators

In this particularly morbid variety we may find astronomers, naturalists, chemists, biologists, and physicians who can be recognized by the following symptom: they love the study of nature but only for its aesthetic qualities—the sublime spectacles, the beautiful forms, the splendid colors, and the graceful structures. If the dilettante is a botanist, he will be anchored forever in the wonder of algae, and especially the diatoms, whose elegant shells capture his admiration. In his fetishistic worship, days pass examining and photographing these interesting creatures in a thousand different ways, arranging them into symbols, fretwork, escutcheons, and other ornamental designs. However, he will never add a new variety to the overflowing catalog of known species, or contribute in the slightest way to our knowledge of the structure, development, and function of these microorganisms.

If the sybarite researcher is a histologist, he will dedicate himself with zeal to the art of producing flashy staining patterns for cells and organic tissues. He will handle the microinjection syringe with ease, and in his naive admiration for the picturesque he will pass his evenings tracing the elegant little networks that carmine and Prussian blue embroider into the capillaries of the intestines, muscles, and glands. He will have mastered completely the most artistic histological staining techniques without ever feeling the slightest temptation to apply them to a new problem, or to the solution of a hotly contested issue.

If he is a geologist, he will be completely engrossed in observing the vivid colors produced in sections of rock by polarized light; if a bacteriologist, he will develop a delight in collecting and cultivating the various chromogenic and phosphorescent microbes; and if an astronomer, he will devote his leisure moments to photographing the mountains on the moon or the spots on the sun.

Why go on? Everyone reading this will recall interesting varieties of this type. They are as likable for their juvenile enthusiasm and piquant and winning speech as they are ineffective in making any real scientific progress.

Bibliophiles and Polyglots

Just as the expert in photomicrography amuses himself with diatoms, or the zoologist with insects, shells, and birds of gorgeous plumage, the bibliophile takes pleasure in reading the newest book or monograph that is "highly important and thought-provoking" but that no one else can seem to find a copy of. Our model of erudition uses this strategy in a marvelous way to amaze his friends.

The symptoms of this disease include encyclopedic tendencies; the mastery of numerous languages, some totally useless; exclusive subscription to highly specialized journals; the acquisition of all the latest books to appear in the bookseller's showcases; assiduous reading of everything that is important to know, especially when it interests very few; unconquerable laziness where writing is concerned; and an aversion to the seminar and laboratory.

Naturally, our bookworm lives in and for his library, which is monumental and overflowing. There he receives his following, charming them with pleasant, sparkling, and varied conversation—usually begun with a question something like: "Have you read So-and-so's book? (An American, German, Russian, or Scandinavian name is inserted here.) Are you acquainted with Such-and such's surprising theory?" And without listening to the reply, the erudite one expounds with warm eloquence some wild and audacious proposal with no basis in reality and endurable only in the context of a chat about spiritual matters.

Discussing everything—squandering and misusing their keen intellects—these indolent men of science ignore a very simple and very human fact. They are censured by their own friends, who feel more pity than respect. They seem only vaguely aware at best of the well-known platitude that erudition has very little value when it does not reflect the preparation and results of sustained personal achievement. All of the bibliophile's fondest hopes are concentrated on projecting an image of genius infused with culture. He never stops to think that only the most inspired effort can liberate the scholar from oblivion and injustice.

Fortunately, we needn't dwell at length on this point in order to correct mistaken social values. No one would deny the fact that he who knows and acts is the one who counts, not he who knows and falls asleep. We render a tribute of respect to those who add original work to a library, and withhold it from those who carry a library around in their head. If one is to become a mere phonograph, it is hardly worth the effort of complicating cerebral organization with study and reflection. Our neurons must be used for more substantial things. Not only to know but also to transform knowledge; not only to experience but also to construct—this is the standard for the genuine man of science to follow.

Thus, let us offer tribute and gratitude to those who leave a wake of brilliant observations, and let us forget those who wore themselves out with nothing to show for it but the transformation of their effusive, sonorous words into phonograph records. Like the popular tenor, the eloquent fount of erudition may undoubtedly receive enthusiastic plaudits throughout life in the warm intimacy of social gatherings, but he waits in vain for acclamation from the great theater of the world. The wise man's public lives far away, or does not yet exist; it reads instead of listens; it is so austere and correct that recognition with gratitude and respect is only extended to new facts that are placed in circulation on the cultural market.

Megalomaniacs

People with this type of failure are characterized by noble and winning traits. They study a great deal, but love personal activities as well. They worship action and have mastered the techniques needed for their research. They are filled with sincere patriotism and long for the personal and national fame that comes with admirable conquests.

Yet their eagerness is rendered sterile by a fatal flaw. While they are confirmed gradualists in theory, they turn out to rely on luck in practice. As if believing in miracles, they want to start their careers with an extraordinary achievement. Perhaps they recall that Hertz, Mayer, Schwann, Roentgen, and Curie began their scientific careers with a great discovery, and aspire to jump from foot soldier to general in their first battle. They end up spending their lives planning and plotting, constructing and correcting, always submerged in feverish activity, always revising, hatching the great embryonic work—the outstanding, sweeping contribution. And, as the years go, by expectation fades, rivals whisper, and friends stretch their imaginations to justify the great man's silence. Meanwhile, important monographs are raining down abroad on the subjects they have so painstakingly explored, fondled, and worn to a thread. And alas, these monographs rob from our ambitious investigator the cherished goal of priority, forcing him to change course. Without losing faith, the megalomaniac takes on another problem, and when he has just about finished the imposing new monument, rivals with scientific contributions extending to the finest detail elicit bitterness again. Finally, he reaches old age amid the indulgent silence of his pupils and ironic smiles of the wise.

All of this happens because when they started out these men did not follow with humility and modesty a law of nature that is the essence of good sense: Tackle small problems first, so that if success smiles and strength increases one may then undertake the great feats of investigation. This cautious approach may not always lead to fame, but at least it will earn for us the esteem of the learned and the respect and consideration of our colleagues.

The dreamers who are reminiscent of the conversationalists of old might be seen as a variety of megalomaniac. They are easily distinguished by their effervescence and by a profusion of ideas and plans of attack. Their optimistic eyes see everything through rose-colored glasses. They are confident that, once accepted, fruits of their initiative will open broad horizons in science, and yield invaluable practical results as well. There is only one minor drawback, which is deplorable—none of their undertakings are ever completed. All come to an untimely end, sometimes through lack of resources, and sometimes through lack of a proper environment, but usually because there were not enough able assistants to carry out the great work, or because certain organizations or governments were not sufficiently civilized and enlightened to encourage and fund it.

The truth is that dreamers do not work hard enough; they lack perseverance. As Gracián has so aptly pointed out in his *Oráculo Manual:* "Some people spend all at the start and finish nothing; they invent but do not progress; everything stops short of completion...The discerning should kill the prey, not spend all of his energy provoking it."

Instrument Addicts

This rather unimportant variety of ineffectualist can be recognized immediately by a sort of fetishistic worship of research instruments. They are as fascinated by the gleam of metal as the lark is with its own reflection in a mirror. They lovingly care for the objects of their idolatry, which are kept as polished as mirrors and as beautifully displayed as images in a cathedral. Peace and monastic discipline reign in their laboratories, where not a spot is to be found and not the slightest noise is to be heard.

Keys jangle incessantly in the ample pockets of the instrument addict. When the professor is not around, it is impossible for assistants and students to access essential monographs and pieces of equipment. Microscopes, spectroscopes, analytical balances, reagents—everything is kept under lock and key. All an assistant would have to do to receive a sentence of doom from the chief would be to damage a Zeiss eyepiece, the refractometer, or the polarizing apparatus. It would be horrible! Furthermore, isn't the instrument addict usually given primary responsibility for laboratory supplies—the inviolable repository of the university? Will the time not come for a strict accounting to his superiors? Investigate? Prove? He will do it some day when he has the time—as soon as the latest monographs containing indispensable information arrive and are consulted! If the government should happen to increase his allotment of supplies, perhaps he could give up part of the hallowed trust for teaching purposes. But in the meantime?

These teachers—and we all remember more than one example—have chosen the wrong profession.[1] They think of themselves as inspiring and zealous officials, when they are in fact simply good housekeepers. Don't they remind one of those excellent housewives who primly set their front rooms in order, keep the furniture scrupulously arranged, polish the floors daily, and receive their relatives and friends in the dining room to avoid dust and disorder?

Obviously, cold-hearted instrument addicts cannot make themselves useful. They suffer from an almost incurable disease, especially when it is associated (as it commonly is) with a distinctive moral condition that is rarely admitted—a selfish and disagreeable obsession with preventing others from working because they personally do not know how, or don't want, to work.

Misfits

There would be many fewer examples of a strange contradiction between genuine vocation and official business, between working for pay and scholarly activity, if a professorship were not so often used merely as a steppingstone to politics, or as advertising to help build a lucrative medical practice. Instead, our professorial candidates should be required to present objective (and in a sense, predictive) evidence of aptitude and suitability through competitive examination.

"One reason for England's prosperity," a Cambridge professor once told me, "lies in the fact that each one of us fills our own post." With certain noble exceptions, the exact opposite occurs in Spain, where many people seem to occupy the same post—not to discharge the responsibilities it carries, but simply to collect the salary, and to enjoy the incidental pleasure of excluding the competent. Who can't think of generals born to be ordinary government officials or justices of the peace, professors of medicine cultivating literature or archeology, engineers writing melodramas, pathologists dedicated to the science of ethics, and metaphysicians sworn to politics? The result of this situation is that instead of devoting all of our spiritual energy to our official duties, we devote only a small part—and that reluctantly, as if it were a painful duty.

However, we would certainly not recommend that the life of the professor, or the man of science in general, should be so austere and strict that his entire life is devoted to professional duties. Instead, we would only hope that whatever energy he has left is spent on light, agreeable pastimes—those perfectly legitimate wanderings of attention that are fueled by the intensity and monotony of daily work.

Some might think that instead of being abnormal, misfits are simply unfortunate individuals who have had work unsuited to their natural aptitudes imposed on them by adverse circumstances. When everything is said and done, however, these failures still fall in the category of abulics because they lack the energy to change their course, and in the end fail to reconcile calling and profession.

It appears to us that misfits are hopelessly ill. On the other hand, this certainly does not apply to the young men whose course has been swayed by family pressure or the tyrannies of their social environment, and who thus find themselves bound to a line of work by force. With their minds still flexible, they would do well to change course as soon as favorable winds blow. Even those toiling in a branch of science they do not enjoy—living as if banished from the beloved country of their ideals—can redeem themselves and work productively. They must generate the determination to reach for lofty goals, to seek an agreeable line of work—which suits their talents—that they can do well and to which they can devote a great deal of energy. Is there any branch of science that lacks at least one delightful oasis where one's intellect can find useful employment and complete satisfaction?

Theorists

There are highly cultivated, wonderfully endowed minds whose wills suffer from a particular form of lethargy, which is all the more serious because it is not apparent to them and is usually not thought of as being particularly important. Its undeniable symptoms include a facility for exposition, a creative and restless imagination, an aversion to the laboratory, and an indomitable dislike for concrete science and seemingly unimportant data. They claim to view things on a grand scale; they live in the clouds. They prefer the book to the monograph, brilliant and audacious hypotheses to classic but sound concepts. When faced with a difficult problem, they feel an irresistible urge to formulate a theory rather than to question nature. As soon as they happen to notice a slight, half-hidden, analogy between two phenomena, or succeed in fitting some new data or other into the framework of a general theory—whether true or false—they dance for joy and genuinely believe that they are the most admirable of reformers. The method is legitimate in principle, but they abuse it by falling into the pit of viewing things from a single perspective. The essential thing for them is the beauty of the concept. It matters very little whether the concept itself is based on thin air, so long as it is beautiful and ingenious, well-thought-out and symmetrical.

As might be expected, disappointments plague the theorist. Current scientific methods are so inadequate for the generation of theories that even those with true genius need to devote themselves to years of struggle and incessant experimental work. So many apparently immutable doctrines have fallen!

Basically, the theorist is a lazy person masquerading as a diligent one. He unconsciously obeys the law of minimum effort because it is easier to fashion a theory than to discover a phenomenon.

Liebig was a good judge of these matters, and he penned some fatherly advice to young Gebhard, a promising chemist who was too inclined toward ambitious synthesis: "Don't make hypotheses. They will bring the enmity of the wise upon you. Be concerned with the discovery of new facts. They are the only things of merit that no one disregards. They speak highly in our favor, they can be proved by all intelligent men, and they create friends for us and command the attention and respect of our adversaries."

There is a great deal of truth in what Liebig wrote. Theories definitely present an exceptional danger to the beginner's future. To instruct carries with it a certain pedantic arrogance, a certain flaunting of intellectual superiority that is only pardoned in the savant renowned for a long series of true discoveries. Let us first become useful workmen; we shall see later if it is our fate to become architects.

The reader may be asking whether or not we are being inconsistent in view of what has already been said about the need for hypotheses. One must distinguish between working hypotheses (*Arbeitshypothesen* of Weismann) and scientific theories. The hypothesis is an interpretative questioning of nature. It is an integral part of the investigation because it forms the initial phase, the virtually required antecedent. But to speculate continuously—to theorize just for its own sake, without arriving at an objective analysis of phenomena—is to lose oneself in a kind of philosophical idealism without a solid foundation, to turn one's back on reality.

Let us emphasize again this obvious conclusion: a scholar's positive contribution is measured by the sum of the original data that he contributes. Hypotheses come and go but data remain. Theories desert us, while data defend us. They are our true resources, our real estate, and our best pedigree. In the eternal shifting of things, only they will save us from the ravages of time and from the forgetfulness or injustice of men. To risk everything on the success of one idea is to forget that every fifteen or twenty years theories are replaced or revised. So many apparently conclusive theories in physics, chemistry, geology, and biology have collapsed in the last few decades! On the other hand, the well-established facts of anatomy and physiology and of chemistry and geology, and the laws and equations of astronomy and physics remain—immutable and defying criticism. "Give me a fact," said Carlyle, "and I will prostrate myself before it."

In short, the beginner should devote maximal effort to discovering original facts by making precise observations, carrying out useful experiments, and providing accurate descriptions. He will use hypotheses as inspiration during the planning stage of an investigation, and for stimulating new fields of investigation. If, in spite of everything, he feels compelled to create vast scientific generalizations, let him do so later on when the abundant observations he has reaped have earned for him a solid reputation. Then and only then will he be listened to with respect and discussed without ridicule. And if fortune smiles, he will someday wear the double crown of investigator and philosopher.

We have now described the major types of failures, highlighting their ethical weaknesses and intellectual poverty in rather bold colors perhaps. We have done this to put them in front of a mirror where they, along with their students and admirers, can observe their defects. We do realize, though, that our diagnoses will do little if any good for the adult and the callous. Instead, our advice is directed to the young who openly crave prestige, even when based on questionable foundations. But even more so, it is directed to those cultured professors who are capable of producing worthwhile results but, with the discouragements accompanying their work, begin to feel the unhealthy and unpatriotic desire to imitate our fruitless braggarts—whether they have been influenced by poor example or lack inner discipline.

If none of the advice in this chapter seems to help those for whom it is intended, they should examine their conscience and decide whether or not they would benefit from undergoing a spiritual cure abroad. The laboratory of a scholar is an ideal sanatorium for wandering attention and a faltering will. Here, old prejudices vanish and new contagions that are both enlightening and sublime are contracted Working beside an industrious and gifted scholar, he who is lacking in will power can receive the baptism of fire in research. In such a laboratory he will observe with commendable envy the fervent ambition to wrest secrets from the unknown; he will absorb the unrelenting scorn toward vain theories and rhetorical discourse; and finally, on foreign soil, he will experience the rebirth of a growing patriotism. And once started down the road of his own work, he now has a store of respectable discoveries to his credit. Back in his native country, he will have learned how to focus his interests, and will now look with disdain—if not pity—on his old idols.

Note

1. We know some who are not content with locking the cabinets in their laboratories; they padlock and seal them before leaving.

6 Social Factors Beneficial to Scientific Work

Material support. Having a profession and doing research work are compatible. The investigator and his family

Like all mental activities, the accomplishments of the scientist are heavily influenced by the physical and moral environments around him. It has been said, with good reason, that the man of learning is like a delicate plant that only thrives in a special medium—soil deposited by the culture of centuries and tilled by society's care and esteem. In favorable surroundings, even the backward type has a feeling of accomplishment, whereas in a hostile or indifferent environment even the sharpest mind is discouraged. How can we go on when no one is interested in our work? Only the stern and heroic have the strength to overcome adverse environmental conditions and wait in obscure resignation for the approval of posterity. But society must not count on heroes because there may be no opportunity for them to appear. Instead, we must rely on people with average skills and ordinary talents who are inspired with a noble patriotism and clear ambition. Governments and educational institutions must contribute to the formation and cultivation of these laboratory patriots by creating a nurturing social environment for them, and by freeing them, insofar as possible, from the preoccupations of material existence.

There is no doubt that for some time to come, and for reasons that will be dealt with later, scientific investigation in Spain will be an exercise in self-denial and sacrifice. Nevertheless, it must be acknowledged that the physical and moral limitations hindering scientific work have been greatly exaggerated. Our university Jeremiahs deplore, and at times with good reason, the lack of financial support; however, more often than not they are being overly dramatic, adopting forlorn rhetorical poses, and even hinting at persecution.

Let us be honest enough to admit that the majority of such dispiriting remarks amount to allegations of *dolce far niente*, or excuses for an absence of loyalty. "I don't have a laboratory and am in a profession that is incompatible with the leisure time necessary for scientific work. Family obligations rob me of the time and money needed for scientific research," and on and on.

It is easy to reduce such lamentations to face value by emphasizing in passing this fundamental truth: *In scientific work, means are virtually nothing whereas the person is almost everything.*

Material Support

A lack of material support is the convenient excuse that many professors, and more than a few physicians, use immediately when asked about their work (when teaching is foreign to them but they are well suited for research). If the grumbler is a *philosopher, lawyer,* or *man of letters,* for example, he will claim a lack of assistants, and above all the absence of a library with specialized journals; if a *bacteriologist, histologist,* or *naturalist,* he will lament the absence of a good microscope, reagents, adequate facilities, and so on; if a *physicist, chemist,* or *engineer,* he will repeat the same refrain, deploring the inadequacy of equipment and laboratory space; if an *astronomer,* he will neglect his work until the government supplies him with the best telescopes; and so on. In short, all of them will agree that our government officials—most of whom are lawyers and litterateurs—have nothing but disdain for experimental science and objective teaching. And taking a familiar stance, the grumblers will not hesitate to place most of the blame for our backwardness upon them.[1]

It would be silly to ignore the fact that often we have had to put up with the old-fashioned rhetoric of certain politicians with no orientation toward Europe, and thus no hope of promoting the intellectual awakening of our country. However, such government officials—focused on ways of the past, and devotees of tradition who are suspicious of modern culture—have all but disappeared.

Today's statesmen undoubtedly have limitations, one of which is not realizing (or at least not advocating) that the greatness and might of nations are products of science, and that justice, order, and good laws are important but secondary factors in prosperity. In any event, they will not make the unpatriotic blunder of denying protection and subsidies to luminaries of the lecture hall and to undisputed scientific competence. It is painful to admit, however, that their naive optimism has also accomplished something else—they have created excellent laboratories for the benefit of young men whose aptitude and loyalty seem rather questionable. And if sinecures are granted to those versed in intrigue and courting favor, and are accompanied by extravagant compensation, how can these benefits be denied to illustrious teachers that have made recognized discoveries or produced scientific work of value?

Although government officials have their weaknesses, they are also capable of noble acts, especially when displaying skill and lively initiative. Nevertheless, these are the same statesmen—whose wills crumble under the demands of friendship, or their political constituencies—who are usually the most willing to reward outstanding merit.

Of course, the facilities just mentioned should be distributed preferentially to the best professors with unquestionable authority. Beginners eager for reputation will meet with greater obstacles. They should not, however, become discouraged. To advance in their noble calling, they must choose between sacrifice and subordination, that is, between a laboratory of their own and a government facility.

In the complete absence of physical resources, any beginner would have to resort to a government laboratory. He will succeed, with determination, to be one of those closest to the master. And as soon as his efficiency and scientific preparation are adequate, what professor would deny him a workbench and paternal advice?

Nevertheless, it would be more pleasing to see the beginner (if physical resources are available) start his apprenticeship in a laboratory of his own—organized and maintained with his own modest savings. There is no doubt that the official establishment, with its master, offers valuable, and in many cases, irreplaceable guidance to us. But in general the work here is subject to many disadvantages. The number of hours worked tends to be short, there is endless conversation and hubbub, students and assistants come and go, and analytical instruments are in constant use. Besides wasting time, these and other annoyances associated with a university laboratory distract attention in a way that is rather harmful to scientific research.

Under such conditions it is much better to work alone, especially if guidance leaves something to be desired. Let books be our masters—wise mentors, serene, no bad temper, and no momentary lapses in ability. We shall happily give them our major commitment—which is to discover ourselves before discovering scientific truth, to mold ourselves before molding nature. To fashion a strong brain, an original mind that is ours alone—this is the preliminary work that is absolutely essential. Then, after achieving technical maturity, what scope and facilities are available to us! Ibsen put these words in the mouth of one of his characters, directed at a friend: "Be yourself!" There is no better way of achieving this than by working alone.

Oh comforting solitude, how favorable thou art to original thought! How satisfying and rewarding are the long winter evenings spent in the *private laboratory,* at the very time when educational centers are closed to their workers! Such evenings free us from poorly thought out improvisations, strengthen our patience, and refine our powers of observation. What care we zealously lavish on our own instruments—each one representing a vanity disowned or a bad habit unindulged! Because we love them we appreciate their fine points, we are aware of their defects, and we avoid the traps they occasionally set for us. In short, we understand their friendly soul, which always responds humbly and quickly to our needs.

But, the reader will remind us, research laboratories are always expensive. This is a sad mistake. It costs little to obtain the necessary tools. The professors, naturalists, physicians, pharmacists, or any others for whom the initial cost and upkeep of a private laboratory for experimental work seems insurmountable would have to be wretchedly poor.

Let me be so bold as to cite myself as an example. With the limited means at the disposal of a lecturer in the provinces, and without any more extraordinary source of additional income than some private tutoring, I established and maintained for fifteen years a laboratory for microscopic work, and an adequate library of journals. My first microscope—an excellent Verick—was obtained on credit. And my experience is not unusual. The thing to do is to launch one's own work on a small scale, but with one's own means; the results are especially educational and fruitful. It is quite well known that the majority of discoveries in physiology, histology, and bacteriology were made by young enthusiasts without reputation or fortune, working in garrets and barn lofts. The institutional laboratory, convenient and richly appointed as it may be, came later as the reward for scientific success.

We could cite dozens of classic examples of modest beginnings. Faraday was an apprentice to a bookbinder when he was carried away by an enthusiasm for science and joined Davy's laboratory as a helper or mechanic. On leaving the laboratory (and not having followed a path to any particular career), he set up a research center that produced a host of admirable conquests—breathing new life into the science of electricity. The great Berzelius began his chemical discoveries in the workshop of his drugstore. Many of the greatest astronomers explored the sky from the rooftops of their homes, armed with mediocre telescopes. Goldschmidt provides a nice example. From the window of his room, aided only by a very small refracting telescope (105 mm), patience allowed him to discover a great many asteroids.

To sum things up, there is poverty of will, rather than lack of means. Enthusiasm and perseverance work miracles. It is the exception when an inexperienced investigator succeeds in launching his career with a memorable scientific achievement from a luxurious and well-appointed laboratory maintained by the state. From the point of view of actual success, it is not the instruments that are costly and require the most time, work, and patience—as we have already pointed out, it is the development and maturing of talent. At the very worst, limited means condemn us to limit our initial steps, to narrow the scope of investigation. But after all, isn't this an advantage?

It is appropriate to distinguish two sciences from this perspective. One is costly and aristocratic, and demands sumptuous temples and rich offerings; the other is more reasonable and familiar, more democratic, and accessible to the most humble purses. And this humble Minerva has special advantages. Its kindness affords better shelter for the blossoming of deep meditation than for the showy and ornate offerings placed upon the altars. Furthermore, there is a noble pride in accomplishment with simple means—the pride of neatness and frugality. Nothing highlights the energetic personality of the investigator better, distinguishing him from the throng of automatons in science, than those discoveries where perseverance and logic get the upper hand over mechanics, where the brain is paramount and material facilities are negligible.

In terms of instruments, the newcomer to *botany, comparative anatomy, histology,* and *embryology* needs a good medium-priced Zeiss microscope with an Abbé condenser, an immersion objective, two dry objectives, and one pair of eyepieces; a small microtome of the Reichert or Schanze type; and some reagents and stains.

The *bacteriologist* and *pathological anatomist* need more varied and expensive equipment, although it is still within reach of the newly established physician or naturalist: a microscope such as the one mentioned above, two burners (one for constant temperatures, and another for sterilizing), test tubes, flasks, animal cages, and so on.

The *physiologist* can begin work with a case of surgical instruments, an apparatus for the animals, a Marey registering cylinder, induction coil, electric batteries, and so on.

The *zoologist,* the *geologist,* and especially the beginner in *comparative and experimental psychology* will be able to satisfy their needs with even less equipment. Nothing is more economical—and intriguing for an even modestly philosophical mind—than the study of instincts, that is, the way animals react in the presence of stimuli or to changes in the environment (the effects of variation, heredity, mutations *per saltum,* and so on); in short, the focus of the classic observations and experiments of Fabre, Réaumur, Huber, Lubbock, Forel, Perrier, Bohm, and others.

The cultivation of *physics* and *chemistry* definitely imposes greater sacrifices. They often require an institutional laboratory that is well stocked with costly analytical equipment and with powerful energy generators. Even so, if our young physicist can learn to confine his interests to special topics under the rubrics of electricity, light, radioactivity, magnetism, and so on, he can also work effectively at home with the aid of a few instruments, making worthwhile contributions and a reputation.

The rule of limiting yourself to one or only a few topics has practical value. Those ambitious to explore the total expanse of a particular science (if that were possible today) would need access to an arsenal of varied and therefore extremely costly instruments, not to mention an extensive workshop. Here is another disadvantage of the encyclopedic mania that we have dissected in earlier chapters.

Having a Profession and Doing Research Work Are Compatible

It is hardly necessary to go out of our way to show that having a profession and doing scientific work as well are not mutually exclusive—they complement and mutually illuminate one another. For those who love observation, professional practice constitutes the best ally of the laboratory. Practice furnishes the material to be studied, and in exchange the laboratory provides theoretical frameworks and practical solutions to the exercise of the practice.

Suppose the careerist is a physician with a typical practice. We strongly believe that he cannot fulfill his mission conscientiously without the aid of a private or institutional laboratory, where he personally works on solving the difficult problems of the clinic with the aid of a microscope and chemical techniques. It is unworthy of him to allege that there is no time, that such work should be carried out by microscopic and chemical laboratories directed by specialists (for the analysis of blood, urine, tumors, microbes, and so on). These laboratories undoubtedly render a useful service, but they are maximally efficient only when the director blends the dual qualities of technician and clinician.

We are certainly not denying that a division of labor is advantageous. But it should be acknowledged that there are certain disadvantages when scientific work becomes too dispersed. One problem is that things that are simply inseparable become separated—that is, tasks that are logically interconnected are assigned to distinct and separate minds. Experimental data and medical judgments that are separated hardly benefit each other, whereas they illuminate and enrich each other when associated in the same intellect.

To be more specific, the following question arises. Physicians invariably acquire a certain amount of experimental skill and mastery of analytical methods if they are inclined to explain the practical aspects of their profession. Under these circumstances, what harm would it do to go a step further and devote himself to original scientific research, without giving up his practice? That this is possible, practical, and fairly simple is proved by the experience of many practicing physicians abroad. Amid the anxieties and pressures associated with the exercise of their profession, and inspired by noble ideals, they have succeeded in organizing private laboratories and have brought honor to themselves and their countries through valuable biological discoveries. From among thousands, let us mention Virchow, who wrote his celebrated work on cellular pathology while a physician in Frankfurt; Robert Koch, a practicing physician in Potsdam, whose research breathed new life into bacteriology through fruitful technical advances and excellent observations; and the brilliant Pleiades of neurologists in Frankfurt (not a university city), where men such as Weigert, Ehrlich, and Edinger devised valuable research methods for histology.

The Investigator and His Family

It is well known that the anxieties and expense involved in raising a family, which is almost impossible on the miserly stipend that the state awards to members of the teaching profession, are one of the excuses used by many of our professors for deserting the laboratory and dedicating their energies to more lucrative pursuits. "Science and a family," they assert, "are incompatible. Because the daily schedule of a professor in research work is so completely absorbing," they add, "how is it possible to ask anyone to share it? The scholar must therefore choose between his intellectual family and his real family, between his ideas and his children."

One must admit that there is an element of truth in these exaggerations. Anxieties at home drain moral and physical strength from the work of research. The ideal university would be a monastery with monks consecrated for life to the study of nature, with few religious obligations to distract them.

We are too imperfect to dedicate ourselves to dual noble causes with equal fervor. He who is eager for heaven is disinterested in this earth. It is well known that psychologists immersed in contemplating the mind overlook the brain. Those who are preoccupied with the material pleasures of everyday life scoff at the microbe. And aspiring for eternal glory makes us less interested in human fame. Fame! It is vain illusion, no doubt, but it is capable of moving mountains and of impelling humanity toward truth and goodness. Like patriotism, the passion for reputation must only be hinted at, never analyzed.

Nevertheless, the austere life of a monk would be an intolerable sacrifice for the majority of scholars. It would appear that the ideal of an intimate sharing of life and aspirations was realized in the famous school of Alexandria. However, those celebrated geometricians and astronomers were undoubtedly married. If woman is evil, grant that she is a necessary one. There are very few austere souls who would look on the fair half of the human race as an unusually decorative specimen in an ornithological collection. Moreover, in trying to win followers, it would be very poor psychology indeed to offer them abstention and martyrdom. Let people deny themselves, but do not impose denial on anyone.

This is an issue that requires protection from the state. It is a purely economic problem. The state has a sacred obligation to make scientific work compatible with a comfortable family life, thus sparing the investigator from painful sacrifices. Like all citizens who believe firmly in the public welfare, scientists should be in a position to satisfy their social instincts fully. In more advanced countries where it is known full well that national prosperity is the fruit of science, this economic problem was resolved satisfactorily years ago. And in Germany and England they have done more. They have converted the lecture hall and laboratory into rich sinecures in their generosity toward professors. The signature of the learned has come to have as much authority in the scientific book as in the checkbook.

What Liebig wrote to Gebhard always comes true in those happy countries: "Set your aims high, and in the end honors and riches will come your way without your having to take the trouble to seek them."

We are still very far from this economic ideal in Spain, although we are moving toward it. As pointed out earlier, it is well known that the physical conditions associated with our professorships, and with laboratory devotees in general, have improved a great deal thanks to laudable government initiatives.[2]

But even if the state were deaf to our appeals, we must not let our work suffer. Our motto ought to be that of the great financiers: earn enough to satisfy all of our needs, particularly the expensive ones—don't be confined to a life of miserable penny-pinching and cowardly abstention.

Let us imagine a worst-case scenario and see how the young investigator can serve his family and his projects at the same time. I take it for granted that our teacher lives in a dreary town in the countryside, without the possibility of a practice. Thus, he lacks the resources needed to satisfy jointly the unquestionable requirements of the home and his beloved research.

Will he deprive himself of everything on the altar of his profession? Will he live a solitary life by renouncing marriage? Of course not. He must serve both his ideals and his proper instincts with equal devotion. As far as his work is concerned, let him turn to inexpensive research that requires few supplies and little equipment but a great deal of effort, and let him use his spare energy on educational projects most directly related to the target of his real love: the textbook, popularizing certain topics, skilled analyses, and finally, private tutoring. The extra income from these sources will allow him to approach his truly noteworthy goals without giving up legitimate expansion of his home. And let him patiently await better times. If his work has true merit, the coveted reward will surprise him in his remote corner. Material success and the tribute of an enviable reputation will be added to the supreme satisfaction that accompanies the rigorous fulfillment of one's duty.

Although many would disagree, we believe that the man of science should be married and should face the pressures and responsibilities of family life courageously.

He will not emulate the selfishness of Epicurus, who did not marry in an attempt to avoid cares and woes, nor the exaggerated refinement of Napoleon, whose only use for a woman was a nurse in old age.[3] For the man of science, the aid of a wife is just as necessary in youth as in old age. A woman at one's side may be likened to a knapsack in battle: without the accessory one fights unencumbered, but after the battle, then what?

At this point we shall make only two suggestions: that the scholar bear in mind his own psychology[4] before selecting a partner, and above all that he avoid at any cost letting others choose her for him. It doesn't take much to justify the marriage of a scholar. In a normal and vigorous young man, bachelorhood is usually an invitation to permanent distraction, if not total abandon to libertinism. One might say that ideas are blossoms of virtue that fail to open their petals and wilt quickly in the fumes of boisterous partying. Moreover, the bachelor lives in a state of complete social preoccupation, where the intrigues of gallantry disrupt the progress of thought far too often. And, as is well known, there is no surer way for a man to prevent a woman from dominating his thoughts than to make her his own. It is common knowledge that a happy home removes selfishness from the soul, ennobles the social instinct, stimulates lofty goals, and fortifies patriotism.

The selection of a partner! Now we touch on a delicate point. What qualities should grace the young woman chosen by the man of science? This is an extremely serious question because it is undoubtedly true that the moral qualities of the wife are a decisive factor in the success of scientific work. Many are under the shadow of wives for whom they are unsuited—and at times society and even humanity as a whole suffers because of the scholar's wife. So many important projects have been interrupted by the selfishness of the young wife! So many careers have been thwarted because of feminine vanity or capriciousness! So many illustrious professors have bowed their necks under the weight of the matrimonial yoke, becoming common seekers of gold, humbling themselves and emasculating their gifts with an insatiable desire to gather as many honors and awards as possible.[5]

When excessive, even the most human and noble impulses of the wife become formidable enemies of scientific work. As is well known, women show an inclination toward a family—a sound tendency for the physical preservation of the race. What a lofty egotism, for it represents the supreme interest of the species! Not without reason and depth, Renan has said: "God loves what a woman loves." She concentrates her love and self-denial on her offspring. Less exclusive, a man can divide his affections between family and society. A woman loves tradition, adores privilege, pays little attention to justice, and is usually indifferent to all work related to change and progress. In contrast, a man truly worthy of the title *Homo socialis* loathes routine and privilege, reveres justice, and in many cases places the cause of humanity above the interests of his family. Because of this, the mother is eager to live only in the memory of her children, whereas the father aspires to survive in the annals of history as well.

Both tendencies—centripetal and centrifugal, contraction and expansion—are legitimate and necessary. The prosperity of the race, and the advance of civilization, depend on their harmony and use. When the altruistic tendency of the man is excessive, offspring languish. In contrast, when the feminine tendency predominates, the family prospers, but society and the state suffer. A spirit of self-denial and sacrifice should reign in the home of the scholar as well as the honored statesman. However, they should not reach the point of creating a bad environment for the development and education of the children. Even taking the collective interest into account, there is no doubt that endless domestic preoccupations and quarrels embitter the life of the thinker, making it difficult for him to carry out scientific and social work.

To sum things up: As a general rule, we advise the man inclined toward science to seek in the one whom his heart has chosen a compatible psychological profile rather than beauty and wealth. In other words, he should seek feelings, tastes, and tendencies that are to a certain extent complementary to his own. He will not simply choose a woman, but a woman who belongs *to him*, whose best dowry will be a sensitive compliance with his wishes, and a warm and full-hearted acceptance of her husband's view of life.

Having come this far, the reader may want us to abandon the field of generalities and define the type of woman most suited for a man of science. We see no reason not to discuss our views here, with all due reserve and circumspection. And to those who may smile at our condescending to such a task, let us merely point out that a factor such as love, which decides one's life, is not a frivolous matter. Nor should one forget that for the man devoted to study, a woman may be the helium that propels him skyward—or the ballast that forces him to land in the marshes of obscurity during the peak of his flight.

The studious man is accustomed to seeking companionship among middle-class women, where four main types stand out: the intellectual, the rich heiress, the artist, and the professional woman.

Intellectual young women have either a scientific or a literary career, or, carried away by an irresistible talent for study, have managed to acquire a sound and varied general education. Because they are such a very rare species in Spain, there is little hope of finding such a pleasing life partner. This is undoubtedly a sensible conclusion, although the small number of women physicians (with a few exceptions) we have known in societies, laboratories, and salons appear determined to console us for the inaccessibility of the type.

In contrast, this variety of woman is plentiful abroad, where genuine prestige allows the female scholar to stand as a collaborator with her husband, exempt insofar as possible from the whims and frivolities of the feminine temperament. A woman like this—intelligent and poised, overflowing with optimism and courage—is the ideal partner for an investigator. She triumphs in the home and in the heart of the scholar, wearing the triple crown of loving wife, intimate confidante, and industrious collaborator. They, let us repeat, are not uncommon in the venturesome nations to the North.

We have observed with great admiration (and more than a little envy) those happy couples in select laboratories that are eagerly dedicated to the same work, each of them pouring into it the very finest of their mental abilities and technical skills. We shall not dwell on the moving example of the Curies, discoverers of radium. Instead, we shall focus on the rather small circle of our own friends and scientific associates. The images of three admirable couples come to mind. Joseph Dejerine and Augusta Dejerine-Klumpke of Paris are dedicated to studying the normal and pathological anatomy of the cerebrum; Jean and Mme. Nageotte, also of Paris, are involved in histological and neurological research; and finally, the wedded pair Oskar and Cécile Vogt of the Berlin Neurobiological Institute are involved in the vast undertaking of mapping out the architecture of the cerebral cortex, like astronomers who pass their lives absorbed in photographing and cataloging stars and nebulae.

But we must repeat, this *ave fenix*—the serious and discreet woman doctor who assiduously collaborates with her husband—has not yet seen fit to appear on our social horizon, where oddly enough the greatest feminine talents are self-taught and entirely absent from regular university studies. The Spanish scientist must therefore choose from among the other types of woman.

Will he direct his course toward the wealthy woman? This seems very dangerous to us. Accustomed to a life of leisure, luxury, and show, it would be miraculous if these tastes were not communicated to her husband. This happened to the illustrious English physicist Davy, who essentially gave up his brilliant career after tying himself to a wife of noble descent. The best part of his life was consumed in the social world of parties and receptions.

It would be incredibly lucky to run across a rich and famous heiress willing to give up the caprices and vanities of her sex by consecrating her gold to the advancement of science. This type of admirable woman abounds in France and England, although in our own country we have never known a professor devoted to the laboratory for whom his wife's riches were not fatal to his work. If our lips were not sealed by discretion, we could illustrate these pages with vivid examples of how frivolously ostentatious tastes of the wife, or the overweening egotism of the family's mother, have interrupted brilliant careers, forcing the young investigator to exchange study for politics, the microscope for the automobile, and redeeming evenings in the laboratory for useless hours spent at parties or the theater.

But let us not censure these rich women too harshly. They have wonderful hearts, but are victims of their own lack of culture. In the end, the interminable reproaches they use to paralyze the noteworthy initiative of their husbands ("Why work when you have the means to enjoy a life of luxury?" and so on) are excusable because they are inspired by conjugal love. The arrogant heiresses who immediately draw attention to the parasitic state and financial inadequacy of their unhappy husbands are far more disagreeable. Embarrassing daily harangues force him to work like a beast of burden, trying to defray the entire cost of a luxurious lifestyle as filled with vanity as with empty ideals. This happens, of course, because the wife has frittered away her dowry on clothes, jewels, luxurious furniture, and excursions to fashionable spas and beaches!

Will the scholar prefer the woman artist or professional litterateur? With a few noteworthy exceptions, such women create a never-ending stream of difficulties for the man of science. It is sad to admit that as soon as she enjoys the talent and culture usually associated with men, her modest charm is lost, she acquires dominating airs, and she lives in a constant state of exhibiting her cleverness and ability. Women are inclined to be somewhat theatrical, but the litterateur or artist is always on stage. And then their tastes are so masculine and involved. At least the wealthy woman is accustomed to indulging her whims at her own expense. She is not a good friend of books and journals, and is drawn instead to jewelry shops and fashionable stores. But the litterateur glances with equal desire over the show windows of jewels and hats and the samples of booksellers.

So, the professional young woman endowed with physical and mental health remains the only likely and desirable partner in fame and in hardship for our young investigator. She is graced with optimism and good character, her education is adequate enough so that she can understand and encourage her husband, and she has the passion necessary to believe in him and dream of his hour of triumph, which she is convinced will happen. Inclined toward simple pleasures and an enemy of notoriety and exhibitionism, she will center her hopes and pride on the health and happiness of her husband, who will find in his home a pleasant atmosphere favoring the germination and growth of ideas instead of reproaches and resistance. And if fame should smile, its brilliance will surround the two foreheads with a single halo.

Fame! The modest wife deserves it also. In the end, she made it possible to execute the great undertaking thanks to her self-denial. By sacrificing fine clothes and jewels, books and journals were available, and she consoles the scholar in his hours of doubt.

Fortunately, this delightful type of woman is not rare in our middle class. The young man who is searching for her in earnest, but either doesn't find her or doesn't know how to unite her to his destiny with all his heart, is most unfortunate. The test lies in winning her for the common good, in appointing himself her spiritual guide, and in molding her character to appreciate the requirements of a serious life of intense work and severely unassuming modesty. In short, as mentioned before, she becomes a complementary mental organ, absorbed in the small things (if running a home and educating children can be regarded as small) so that the husband, free of anxiety, may occupy himself in the great things—in germinating and feeding his beloved discoveries and scientific hypotheses.

Notes

1. At the present time (1923) some laboratories in Spain are so richly endowed that they are the envy of the greatest scholars in other countries. Yet little or nothing is produced in them. The fact is that our statesmen and educational institutions have forgotten two important things. First, claiming to be an investigator does not make one so; and second, discoveries are made by people, not scientific instruments or overflowing libraries.

2. He who writes these lines, the most humble of Spanish professors, would be guilty of ingratitude if he did not clarify a fact that speaks very highly for the generosity of our government. The mere news by telegraph that the Moscow prize granted by the International Medical Congress in Paris (1900) had been awarded to a Spaniard was enough for me to be sought out instantly in the corner where I toiled in silence, and for an excellent laboratory to be placed at my disposal. Aside from the great honors received from the major scientific organizations in the world, two new offerings of my lucky star obtained later (the Helmholtz medal [in 1905] and the Nobel prize [in 1906]) provided me with the satisfaction of reflecting that the modest sacrifice made by the Spanish state had not been fruitless for science.

Fortunately, my experience is not unique. Every one of our countrymen who has been singled out by foreign science to receive honors and awards has done so without wishing for or seeking them. Therefore, the selfish, who always place reward before merit, should realize that in our country as well, the serious cultivation of science is regarded as good business.

3. The celebrated remark of Bonaparte when speaking to the Council of State is familiar: "If it weren't for a man aging, he would just as soon do without women."

4. We allude here particularly to the effects of mental concentration and intense work, which may end up keeping the scholar perpetually distracted. He may then become as indolent and careless in the education of his children as in the administration of his property.

5. We could cite more than twenty young men with great talent and excellent training whose early attempts at research were shipwrecked on the shoals of matrimony. Currently, most of our best producers are unmarried, especially those in biology.

7   Stages of Scientific Research

Observation. Experimentation. Working hypotheses. Proof

We agree with writers on logic, and in particular with E. Naville, that three successive operations are necessary in all scientific research: observation and experimentation, hypothesis or supposition, and proof. Sometimes the inquiry itself is not based on personal observation, but on a feeling rooted in criticism—an *a priori* dislike for a rather widely held tenet. Nevertheless, it is clear that such feelings are often based on the objective results of personal observation, no matter how superficial, of topics or material related to the problem to be solved.

Observation

The advice given by those who formulate literary precepts has very direct application to scientific research. Perez de Ayala has stated it very wisely and skillfully: "Look at things as if for the very first time." That is, admire them afresh, disregarding what we remember from books, stilted descriptions, and conventional wisdom. We must free our minds of prejudice and fading images, and make a definite point to see and judge for ourselves, as if the object had been created expressly for the gratification and delight of our intellect alone. In short, we must re-create, insofar as possible, the state of mind of the fortunate scholar who discovered the fact under consideration, or who first stated the problem—a blend of surprise, emotion, and lively curiosity.

This is intimately linked with another rule that is insisted upon by those devoted to scientific research. It is not sufficient to examine; it is also necessary to observe and reflect: we should infuse the things we observe with the intensity of our emotions and with a deep sense of affinity. We should make them our own where the heart is concerned, as well as in an intellectual sense. Only then will they surrender their secrets to us, for enthusiasm heightens and refines our perception. As with the lover who discovers new perfections every day in the woman he adores, he who studies an object with an endless sense of pleasure finally discerns interesting details and unusual properties that escape the thoughtless attention of those who work in a routine way.

Descending now to more solid ground, let us formulate some indispensable advice for precise observation in the field of biology.

The work should be carried out under the best possible conditions, and to this end one should utilize the best analytical instruments, as well as research methods that can be used with the greatest confidence. If possible, we should use a variety of methods on the same problem, correcting the weaknesses of some with the revelations afforded by others. Let us select the most accurate technique, which at the same time provides the most conclusive mental images. It is also important to avoid any tendency to hasty judgment in the evaluation of data. We should repeat the experiments in a hundred ways until we are certain that they are absolutely consistent, and until we are certain that we have not been victimized by any of the false paths that lead young explorers astray (especially in microscopic work).

If our study is concerned with an object related to anatomy, natural history, and so on, observation will be accompanied by sketching because, in addition to other advantages, the act of representing something disciplines and strengthens attention. It forces us to examine the entire phenomenon, thus preventing the details that commonly go unnoticed in ordinary observation from escaping our attention. This is all the more true because morphological studies are usually incomprehensible without drawings. The great Cuvier stated with good reason that "without the art of drawing, natural history and anatomy would have been impossible."[1] It is not without reason that all great observers are skillful at drawing.

When we are not entirely satisfied with the results, even though the appropriate technique was used, we must repeat the experiments as many times as necessary to obtain everything that the method can provide. To this end, it is extremely useful to have before us a very good example of the results prepared by the author of the method himself, or by one of his most trusted disciples—because it may be compared directly with our own results. One must bear in mind that new facts are not discovered by the one who first observes them. They are discovered by the one who uses an excellent technique and is able to establish them with a full range of evidence, and in so doing convincing everyone. As we have already pointed out, almost all great scholars in the biological sciences owe their victories to the complete mastery of one or more graphic or experimental techniques.

Experimentation

In many sciences (physiology, pathology, physics, chemistry) experimentation is more important than observation itself. It is impossible to discover anything in physics or physiology without envisioning an original experiment, without subjecting the phenomenon of interest to more or less new conditions. And the study of morphology itself (histology, anatomy, embryology), where observation seems enough, is acquiring a more experimental character every day. We owe this change, of course, to extremely valuable conquests that might never have happened by following the much-traveled road of analyzing static forms anatomically. Among a thousand examples that might be cited, let us recall the production of *artificial parthenogenesis* in the *starfish* (a sexed animal) by substituting an influx of seawater with magnesium chloride for natural fertilization (by sperm); the interesting experiments on *merogony* (destruction of early cells produced by division of the fertilized egg) carried out in amphibians by Roux, Hertwig, Wilson, and others, proving that each original cell has the capacity to generate a complete embryo and thereby clearly disproving the embryological theories of *preformation* and *mosaicism;* the work of Jean Nageotte, Marinesco, and others dealing with the transplantation of nerves and ganglia and proving that the morphology of nerve cells is simply dependent on their surrounding chemical environment; the marvelous results obtained by Harrison and by Carrel and his school (at the Rockefeller Institute) on the artificial culture of cells from normal and pathological tissues, *in vivo* and *in vitro;* the interesting experiments of H. De Vries and many other modern naturalists on the *mutation* of species and the mechanism of heredity; and so on.

Such admirable triumphs should inspire us to complement, insofar as possible, the merely static examination of form with the interventions of the experimental method. In so doing, we induce marked alterations in the normal biological state of cells and organs. Doing this simplifies the logical process of determining causality and physicochemical mechanisms underlying the phenomena of interest. In observation alone, changes in the conditions associated with the phenomenon undoubtedly occur, but they are rare and episodic under natural conditions. By using experimentation, the time scale is shortened, and we become masters of natural determinism, as well as of the reasons for change.

Working Hypotheses

Once data have been gathered, it is time to assess their significance, and to determine relationships with the more general principles of science. When facing an unusual event, the mind's initial response is to formulate a hypothesis that tries to explain the event and place it in the context of known laws. Then an experimental test will render final judgment on whether or not the hypothesis is correct.

Reflecting on the nature of valid hypotheses, one may note that they are usually fortunate generalizations or daring inductions that allow one provisionally to consider the recently discovered data as one instance of a general principle, or as the unknown effect of a known cause. For example, evolution—which has been so useful in the biological sciences—is simply a generalization to all life of the law of heredity, which is demonstrated conclusively in the life history of every species. When Lavoisier devised the theory of animal heat, he in essence reduced the hitherto unknown phenomenon of animal respiration to the general law of heat production by carbon oxidation.

When generating hypotheses, we should bear in mind the following truisms. (1) A hypothesis is necessary; without it phenomena cannot be explained. (2) Hypotheses may be contrasted or compared, or at least verified, at a more or less distant time in the future. In fact, hypotheses that cannot be tested by observation or experiment leave the problem unilluminated. Such hypotheses are only unifying syntheses, mere verbalizations; they do not explain the data. (3) Hypotheses should be easily understood in terms of chemistry and physics. And, if possible, they should lie in the realm of pure mechanism, as Lord Kelvin had wished—hypotheses that are obscure or too abstract run the risk of being empty verbal descriptions. (4) Hypotheses solve quantitative problems in a qualitative way, avoiding occult properties and metaphysical considerations. (5) If possible, hypotheses should also suggest new research and arguments. And if they do not solve the problem, they should at least steer us in the right direction, stimulating new and more accurate concepts (the *working hypothesis* of Weismann). Even if wrong, a hypothesis can be useful provided it is based on new observations and suggests an original path for scientific thought. In any event, an explanation proved false always has an advantage: by exclusion, it restricts the scope of the imagination and eliminates untenable solutions and sources of error. Le Blon says with good reason that "he who refuses to accept hypothesis as a guide is resigned to accept chance as a master."

Many renowned men of learning, and especially the great physicist Tyndall, have dwelt eloquently on the importance of hypothesis in science, and on the significant role played by the imagination in formulating good and useful theories.

And I agree. If the hypothesis is a much abused weapon, it is also an instrument of logic. Not even observation itself, which by its very nature is passive, can be realized without hypotheses. Good or bad, a conjecture (or any attempt whatsoever at explanation) should always be our guide. No one searches without a plan.

Even so-called accidental discoveries typically owe something to a guiding idea not sanctioned by experience, to an idea with the virtue of leading us to poorly explored or uncharted territory. Pardon the common simile, but it is worth repeating that what happens under these circumstances is like what happens when an acquaintance appears in the street among the crowd—at the very moment we are thinking of him. Of course he would have passed by unnoticed if we hadn't been thinking of him. Catalyzed by a hypothesis, we may find something in the data that we were not looking for, but this is better than finding nothing at all—which is precisely what happens to the entirely passive observer of natural phenomena. As Peisse has said, "The eye only sees what it is looking for in things, and it looks only for what lies in the mind in the form of ideas."

One shouldn't need reminding that all great investigators have been prolific hypothesis generators. It has been said with deep conviction that hypotheses are the first murmurings of reason in the darkness of the unknown; the sounding instrument lowered into the mysterious abyss; in short, the high, lofty, and audacious bridge connecting the familiar shore with the unexplored continent.

Hypotheses have been greatly abused. However, it must be admitted that without them our store of firm data would be exceedingly limited, and would grow very slowly. The hypothesis and the objective datum are bound together in a close etiological relationship. Theory is a valuable tool, in addition to its conceptual and explanatory benefits. "The scientist must not forget," affirms Huxley, " that hypotheses must be considered a means, never an end." To observe without thinking is as dangerous as thinking without observing. Hypothesis is our most valuable intellectual tool, and like all tools it may become nicked and rusty, in constant need of repair and replacement. But it would be almost impossible to carve out any shape at all in the hard block of reality without it.

It is difficult to prescribe rules for generating hypotheses. People without a certain intuition for causal linking—a divinatory instinct for perceiving the idea behind the fact and the law behind the phenomenon—will rarely devise a reasonable solution, whatever his gifts as an observer. Nevertheless, it is useful to outline some general views or standards for hypotheses in biology. The following may be useful when thinking about how to formulate such a hypothesis.

1. *Nature uses the same means for equivalent ends.* There are few exceptions to this principle and its use often allows us to compare a phenomenon that we do not understand with one that we do. For example, when *mitosis* or *karyokinesis* was discovered in the large cells of the larval triton and in the salamander, one could reasonably expect to find a similar phenomenon in the cell divisions of humans and the higher vertebrates, either under normal or pathological conditions. And of course the prediction was confirmed by additional work. Let us cite another example: Research by Kuhne, Krause, Ranvier, and others established that vertebrate motor and sensory nerve fibers end freely as varicose arborizations. According to the rule under consideration, it could be anticipated that the same arrangement would be found not only in the nerve centers of vertebrates but in the nerve

centers of invertebrates as well. This reasonable suspicion was later confirmed by us, as well as by Kölliker, Lenhossék, Van Gehuchten, and others, for vertebrates; and by Retzius, Lenhossék, and others, for invertebrates. There is no need to provide further examples.

2. *View the problem in its simplest forms.* Because ontogeny and phylogeny present two almost corresponding series of forms ranging from the simple to the complex, there is no better way to throw light on the structure of a complex, almost baffling organ in the higher vertebrates than to study it in its simplest form, either during the development of the individual or in the evolution of the species. An excellent method for determining the meaning of something is to find out how it comes to be what it is. By marking its place in the evolutionary chain we shed light on its anatomical and physiological properties without consciously meaning to do so.

3. *All natural arrangements, however capricious they may seem, have a function.* This teleological principle applies to all the quirks of plant and animal structure, although it was formulated to deal with vestigial organs. In stating this law, we do not claim that each organ represents a direct incarnation of the Creator, as Linneaus, Cuvier, and Agassiz assumed. We are simply pointing out that for whatever reason, every organ conserved by nature (that is, maintained by heredity for long periods of time) almost always plays a useful role for individuals of the species; superfluous or unfavorable arrangements brought about by change and other conditions are eliminated eventually. According to this principle, we should assume an important function for all organs or tissues that tenaciously persevere throughout the animal series, and a less important function (at least for the life of the individual) for others featured less prominently in the

animal kingdom. Physiologists constantly use and abuse this postulate when trying to interpret the functional dynamics of the circulatory, digestive, locomotor, and other systems. A great deal of light is shed on these dynamics by physics and chemistry, or as Tetamendi said, *the present state of our industrial knowledge.*

There are undoubtedly exceptions to the utilitarian principle just outlined, but they are rare and easily explained by the recent and thus incomplete adaptation of an organ to new conditions (organs atrophied through disuse, and so on). In his *Studies on Human Nature,* Metchnikoff writes brilliantly on these biological incongruities, which are more common in humans than animals. His conclusions are derived from Lamarck's principle on the use and disuse of organs.

It is well known that a hypothesis is always formulated to explain observations that have been made. Without broaching the difficult problem of *scientific explanation* (which would involve arguments inappropriate for this little book), we shall simply point out that when considering natural phenomena the intellect can adopt one of two equally valid approaches: (1) the new observation can be assigned to an established law or principle (Meyerson's *legalistic* explanation); or (2) the new observation can be dealt with logically in terms of pure *mechanism,* and entered humbly into the equations of dynamics. This is in addition to its *legality* or relationship to a general law. For Maxwell and most other modern physicist-philosophers, the second mode of explanation represents a greater level of scientific understanding, and requires the use of general theories that are on a higher plane than empirical laws.

We are forced to admit that our minds imperiously demand understandable theories presented in mechanistic terms. Whatever resists physical interpretation runs the serious risk of being a mere plaything of the imagination without an objective basis in reality. The psychological reason for this escapes us. As Bergson would argue, perhaps it lies in the fact that because our concepts are modeled on discontinuous sensations, the imagination can only forge the ultimate representation of things out of something resembling the sensorial data itself, which involves variations in the movement of discontinuous parts, disturbances in configuration, and the dynamics of physical systems.

In physics, chemistry, and astronomy, hypothetical explanations based on *mechanistic reduction* are very common, and the investigator should derive inspiration from them. They help his ideas remain plastic, and allow his hypothesizing to advance. However, in anatomy, biology, and pathology we shall usually have to remain content with *legalistic hypotheses*. While they may not satisfy our eagerness for comprehension, they are enough to placate those two great desires of reason, to act and to predict.

Proof

Once a hypothesis is clearly formulated, it must be submitted to the ratification of testing. For this, we must choose experiments or observations that are precise, complete, and conclusive. One of the characteristic attributes of a great intellect is the ability to design appropriate experiments. They immediately find ways of solving problems that average scholars only clarify with long and exhausting investigation.

If the hypothesis does not fit the data, it must be rejected mercilessly and another explanation beyond reproach drawn up. Let us subject ourselves to harsh self-criticism that is based on a distrust of ourselves. During the course of proof, we must be just as diligent in seeking data contrary to our hypothesis as we are in ferreting out data that may support it. Let us avoid excessive attachment to our own ideas, which we need to treat as prosecutor, not defense attorney. Even though a tumor is ours, it must be removed. It is far better to correct ourselves than to endure correction by others. Personally, I do not feel the slightest embarrassment in giving up my ideas because I believe that to fall and to rise alone demonstrates strength, whereas to fall and wait for a helping hand indicates weakness.

Furthermore, we must admit our own absurdities whenever someone points them out, and we should act accordingly. Proving that we are driven only by a love of truth, we shall win for our views the consideration and esteem of our superiors.

Excessive self-esteem and pride deprive us of the supreme pleasure of sculpting our own lives; of the incomparable gratification of having improved and conquered ourselves; of refining and perfecting our cerebral machinery—the legacy of heredity. If conceit is ever excusable, it is when the will remodels or re-creates us, acting as it were as a supreme critic.

If our pride resists improvement, let us bear in mind that, whether we like it or not, none of our tricks can slow the triumph of truth, which will probably happen during our lifetime. And the livelier the protestations of self-esteem have been, the more lamentable the situation will be. Some disagreeable character, perhaps even with bad intentions, will undoubtedly arrive on the scene and point out our inconsistency to us. And he will inevitably become enraged if we readily correct ourselves because we will have deprived him of an easy victory at our expense. However, we should reply to him that the duty of the scientist is to adapt continuously to new scientific methods, not become paralyzed by mistakes; that cerebral vigor lies in mobilizing oneself, not in reaching a state of ossification; and that in man's intellectual life, as in the mental life of animals, the harmful thing is not change, but regression and atavism. Change automatically suggests vigor, plasticity, and youth. In contrast, rigidity is synonymous with rest, cerebral lassitude, and paralysis of thought; in other words, fatal inertia—certain harbinger of decrepitude and death.[2] With winning sincerity, a certain scientist once remarked: "I change because I study." It would be even more self-effacing and modest to point out: "I change because others study, and I am fortunate to renew myself."

When the work of confirmation sheds little light, let us conceive new experiments and try to place ourselves in the best position to evaluate the implications of the hypothesis. In anatomy and physiology, for example, it is often impossible to clarify the structure or function of a complex organ because we are attacking the problem from the most difficult angle, trying to solve it in humans or the higher vertebrates. But if we approach the embryo, or lower animals, nature reveals herself more simply and less elusively. Here she offers us an almost schematic plan of the structure and function we are looking for. Quite often, our hypothesis will receive unexpected and definitive confirmation.

In summary, the order of events followed by an investigator in the conquest of scientific truth is usually as follows: (1) observation of the facts demonstrated, based on methods that are decisive, clear, and highly precise; (2) experimentation, which creates new conditions for observing the phenomena; (3) criticism and elimination of erroneous hypotheses, and elaboration of a rational interpretation of the facts—thus subordinating the latter to a general law or principle, and if possible to a physicochemical representation or model; (4) proof of the hypothesis by new observations or repeated experiments; (5) replacement of the hypothesis, if it does not fit the facts, with another hypothesis that is in turn submitted to rigorous objective analysis; and (6) applying the implications of the hypothesis, now regarded as established, to other spheres of knowledge.

Notes

1. This quote was mentioned by the distinguished Professor Pou-Orfila in an excellent pamphlet dealing with the study of anatomy, *Observaciones sobre la enseñanza de la Medicina* (Montevideo, 1906).

2. In politics, the worship of inflexibility or resistance to change is considered a virtue, whereas in science it is almost always an unmistakable sign of pride or shortsightedness. Flexibility is one of the features that best conveys an investigator's honesty. In our view, he who cannot abandon a false concept brands himself as either stupid, dated, or ignorant. Only fools and those who don't read persist in error. Those who insist on being inflexible at all costs seem to declare, with their Olympian disdain for all scientific innovation: "I am worthy, and I know so much that no matter what science discovers, it will not force me to change my views one iota." In essence, the cerebrum is like a young tree whose foliage spreads and develops a more intricate pattern with study and meditation. Therefore, aspiring to avoid change in matters open for discussion is to wish that the future tree will never grow beyond the stage of a sapling or never grow curving, outstretched branches. Science teaches us that man, in his passage through life, renovates himself physically and mentally over and over. During the life of an individual there are many different incarnations that may almost interrupt the continuity of consciousness and the view of one's own personality. New reading and changes in one's ethical and intellectual environment continually alter and improve the atmosphere within, along with clarifying and refining judgment. After fifty years have gone by, who would dare, with any sincerity, to defend all of the concepts his personality held at twenty—that is, the thinking of inexperienced and magnanimous youth?

8 On Writing Scientific Papers

Justification for scientific contributions. Bibliography. Justice and courtesy in decisions. Description of methods. Conclusions. The need for illustrations. Style. The publication of scientific works

Justification for Scientific Contributions

Mr. Billings is a scholarly Washington librarian who is burdened with the task of classifying thousands of publications where essentially the same facts are presented in different ways, or truths known since antiquity are expounded upon. He counsels scientific writers to govern themselves by the following rules: (1) Have something to say, (2) say it, (3) stop once it is said, and (4) give the article a suitable title and order of presentation.

I believe these guidelines are very applicable in Spain, which provides a classic example of hyperbole and pompous exposition. As a matter of fact, the first essential in dealing with scientific matters (when one is not inspired by the mission of teaching) is to have some new observation or useful idea to communicate to others. Nothing is more ridiculous than the presumption of writing on a topic without providing any real clarification—simply to exhibit an overly vivid imagination, or to show off pedantic knowledge with data gathered second- or thirdhand.

When we take up our pens to write a scientific article, let us bear in mind that it will probably be read by some well-known scholar who is too busy to waste time reading things he already knows, or scanning mere rhetorical exercises. Unfortunately, many of our academic lecturers are guilty of this capital offense. Numerous doctoral dissertations and more than a few articles in our professional journals seem to have been written not with the intention of shedding new light on a subject but of displaying the eloquence of the author, who is willing to accomplish the difficult task of writing in any slipshod manner possible—and the longer the better (for they make sure that what isn't taken up in doctrine is taken up in space)—without taking the trouble to think. Let us take note of the many discourses with titles such as the following, which seem to be invented by laziness itself: "The general idea . . . ," "Introduction to the study of . . . ," "General considerations regarding . . . ," "Critical evaluation of the theories of . . . ," and "The importance of such and such a science"—titles providing the author with the priceless advantages of avoiding bibliographic research and disposing of the material however he wishes, without forcing himself to treat anything thoroughly or seriously. This is not meant to detract from certain perfectly conceived and executed articles that appear now and then with titles similar to the ones just mentioned.

Thus, let us assure ourselves, based on a careful literature review, that the facts or ideas we wish to write about are original, and furthermore, let us guard against premature publication of our observations. When our thoughts still waver between various conclusions, and we are not fully conscious of having hit the target squarely in the center—these are signs of having left the laboratory too soon. It would then be wise to return to it and wait until our ideas have crystallized fully.

Bibliography

Before explaining our personal contribution to a field, it is customary to trace the history of the problem. This is done not only to establish a point of departure but also to render the tribute of justice to the scholars who have preceded us and opened the path of investigation for us. When the young investigator is tempted to overlook dates and references—whether for love of brevity or through sheer laziness—let him bear in mind that others will have the chance to repay him in kind by purposely keeping his work obscure. Such behavior is as ungenerous as it is discourteous because the majority of scholars obtain no greater reward for their painstaking work than the esteem and applause of the learned who, as we have noted earlier, are a tiny minority.

Respect for the ownership of ideas is well-practiced only when one comes to be the owner of ideas that travel from book to book, sometimes with the author's name and sometimes without, and sometimes with mistaken paternity. After being victimized by such omissions and unfair silences, one comes to realize that each idea is a scientific creature, and that the author of its existence—the one who gave it life at the cost of great hardship—utters the same outraged cries on seeing his paternity disregarded as would the mother who has had the life she nourished within her snatched away.

If we believe in justice, let us do so in the context of the article as well. Thus, at the time of a discovery, we need to arrange in exact chronological order, and concisely, the list of relevant names or references. For such a list to be organized logically, it must begin with the originator and finish with the confirmers and refiners. First and foremost, careful study of the literature will spare us from committing an injustice, and thus from the inevitable claims of those with priority.

Justice and Courtesy in Decisions

When accounting for historical precedence, it is often necessary to make decisions about the far-reaching effects of work that is not entirely familiar to us. Needless to say, we must work not only with impartiality but also with irreproachable courtesy and with a pleasant and almost flattering style in such appraisals. As indulgent as we are to the mistakes of the beginner, we must be respectful and unassuming with the mistakes of the great names of science. Let us always fear that our judgments may reflect the shallowness of impatience or the mirages of juvenile enthusiasm. Thus, before deciding to reject a commonly accepted fact or interpretation, let us reflect maturely and bear firmly in mind before expressing our objections that if some of the wise are noble and kind, many more have irascible temperaments reflecting the haughtiness of a Caesar, along with a keenly sensitive vanity. The Horatian phrase *genus irritabile vatum* is even more applicable to the wise than to poets. The discerning Gracián once wrote: "Scholars have always been a little touchy; add science and you add impatience."

Using these precautions, we shall avoid as much as possible a continuing scorn for our work and embittering quarrels and debate—which cost peace of mind and time without gaining prestige or authority. In evaluating our merit, only original contributions to science will be considered, not debating skill and nobility.

When unjustly attacked and forced to defend ourselves, let us do so nobly. Unsheathe your sword, but with tip blunted—adorned with a bouquet of flowers, to use a common phrase.

It is painful to admit that in a majority of cases the objectors are defending their own infallibility rather than a doctrine or principle. Eucken very aptly notes that under the pretext of refuting principles, "everyone defends himself in his own way and according to his nature . . . It is the instinct of self-preservation that reacts."

Unfortunately, when forced to deal with this type of opponent (which is inevitable at times because all truth exasperates champions of error), it is naive to hope that they will be convinced. We must look to the public rather than to them. Let us present conclusive proof; let us strengthen our thesis with new objective data; and let us endure personal attacks and polemic ambushes in silence. In such contests it is important to defend the truth before defending ourselves.

So many vexations and annoyances come one's way by forgetting these well-known rules of foresight and discretion! Violent, acrimonious replies and rancorous silence usually indicate a lack of courtesy and kindness in explaining and evaluating the work of others.

Let me cite a few concrete examples to guide and instruct the beginner. Typically, critical examination deals with errors of fact or observation, or with errors of interpretation.

1. *Errors of observation or recognition of fact.* Generally speaking, the learned discuss interpretations, not facts, supposing that the investigator, no matter how modest, does not engage in analysis without sufficient preparation. And that is

precisely why such mistakes are so grave, reflecting as they do a singular lapse of intelligence or inexperience with techniques. However, let us guard against flying into a rage when the absurdity becomes obvious; let us instead be tolerant and bear in mind that even the wisest of scholars can fall into error in moments of distraction or carelessness. Instead of criticizing the inconsistency harshly, let us excuse it with kindness, noting that it deals with very difficult observations, where mistakes are common and almost unavoidable. Let us not ascribe the error to ignorance, but rather to a limitation of the method employed or to the prejudices of the school of thought that inspired the work.

When such excuses appear inappropriate despite the best intentions in the world, let us attribute the blunder to the use of insufficient or inappropriate material, adding that if the author had made use of the same objects of study as ourselves, he undoubtedly would have arrived at the same conclusions because he has more than adequate talent and skill, as demonstrated amply in previous publications. In short, let us try to cheer him up, dwelling at length on the more or less unique attention to detail shown in the work or on how excellent the description is. In fact, we might go even further and compliment the neatness and precision of his drawings. Overall, the major aim of our remarks will be to soften his bitter reaction to the verdict and convey to our adversary the feeling that his eager efforts have not been wholly useless for the progress of science.

2. *Theoretical error.* Let us suppose that the author has interpreted the data incorrectly, and has thus formulated an arbitrary hypothesis that has no basis in observation. The critical verdict will be softened with remarks to the effect: "The explanation offered is certainly too daring. On the other hand, however, it is remarkably ingenious, and sug-

gests that the author has given it deep thought, and that he has a lofty, philosophical mind. It is such a pity that in formulating his concept he did not take into account such-and-such data, which seriously contradict it! In any case, the hypothesis is intriguing and merits serious attention and discussion."

However, the theoretical interpretation may be so trivial and obvious that even offering an excuse for it may seem like flattery. In that case, it is best to let it pass in silence. As in the foregoing case, one should merely note the valid observations (if any), and the literary, philosophical, or pedagogical merit of the work.

Description of Methods

It is also important to outline fully the method or methods of investigation followed by the author, either at the beginning or end of the monograph. Do not emulate those scholars who keep to themselves the technique employed, under the pretext of improving it later. This approach is a throwback to the almost forgotten custom of mathematicians and chemists in centuries past, who kept details of the procedures truth had revealed to them under wraps, inspired by the childish vanity of astonishing people with their powers of insight. Fortunately, this esoteric cult is vanishing from the domain of science, and the mere journal reader of today can know the details and *tours de main* of particular methods almost as well as intimate friends of the discoverer.

Conclusions

Once the observations that are the fruit of our inquiry are explained in a clear, concise, and systematic way, the article should be brought to a close by summarizing briefly the positive data contributed to science—the motivation for our working on the problem in the first place.

A plan that is not followed by all, but one that seems quite laudable to us, involves calling the reader's attention to problems that remain to be solved, so that other observers may lend their efforts and complete our work. In pointing out to successors the direction of further research, and the problems our own work have not been able to solve, we increase the likelihood that our discoveries will receive prompt and full confirmation, and provide a generous boost to young investigators eager for reputation.

The Need for Illustrations

If our research deals with either microscopic or macroscopic morphology, it will be essential to illustrate the descriptions with figures that have been copied very precisely from the material. No matter how exact and minute the verbal description may be, it will always be less clear than a good illustration. This is true because the graphical representation of an object reflects how carefully the observations were done, and sets an invaluable precedent for those who attempt to confirm our assertions. With good reason, almost equal credit is now given to the one who faithfully draws an object for the first time, as compared to the one who makes it known only through verbal description, which may be more or less complete.

If the objects represented are too complex, explanatory or semischematic diagrams should accompany the exact drawings. Finally, in some cases ordinary photography or photomicrography will be very useful. The latter, of course, guarantees the objectivity of our description.

Style

Finally, the style of our work should be genuine, didactic, sober, simple, and free of affectation, and it should reveal a preoccupation with order and clarity. Undue emphasis, oratory, and hyperbole should never enter into purely scientific writing, unless we wish to forfeit the confidence of scholars, who will come to regard us as dreamers or poets, incapable of studying and applying cold logic to a problem. The scientific writer will constantly aspire to reflect objective reality with the perfect serenity and candor of a mirror, drawing with words as a painter with his brush. In short, forsake the pretensions of the stylist and the fatuous ostentation of philosophical depth. Do not forget the well-known maxim of Boileau: "That which is well conceived is stated clearly."

The pomp and display of fine writing may have a place in the popular book, in inaugural addresses, or even in the prologue or introduction to a scientific work; but it must be confessed that excessive rhetoric in a scientific monograph strikes one as a bit strange and ridiculous.

This is true even if we disregard the fact that a mask of rhetoric often lends fuzzy contours to ideas, and the fact that unnecessary comparisons make description verbose—needlessly distracting the attention of the reader, who certainly does not need the frequent use of common images to assimilate ideas. Writers, like lenses, may be divided into two types, *chromatic* and *achromatic.* Dispersions have been corrected completely in the latter, and they are able to concentrate with precision on the ideas gathered through reading or observation, whereas the former lack the brakes of correction and enjoy expanding the contours of ideas with colorful rhetoric and brilliantly hued shadings—achieved entirely at the expense of the force and accuracy of the ideas themselves.

So-called chromatic or dispersive intellects may be quite useful in literature and in oration because the common man—the final judge of artistic work—needs *rhetorical tricks* to accept certain truths. However, in the exposition and discussion of purely scientific topics, the public is a cultured and select body. We will certainly offend its learning and good taste by dealing with problems too *ab ovo* and losing ourselves in declamatory amplification and useless detail. The following maxim of Gracián, which was praised by Schopenhauer, must be our standard: "Whatever is good, if brief is twice as good." He also had the following advice: "One should speak as in a will—fewer words mean less litigation."

The following constitute excellent prophylaxis against what Fray Candil graphically called "rhetorical flatulence," which we regard as a manifestation of superficial sectionalism, and a powerful cause of our scientific backwardness: an extreme focusing of attention, the habit of giving more importance to thought and action than to words, and a conviction that the scientific problem we are studying has not advanced a single step toward solution after the framing of an appropriate image or sentence.

The Publication of Scientific Works

When the investigator enjoys worldwide recognition, he will be able to publish his scientific contributions in any local or foreign journal in his specialty. Scholars interested in the topic will not be deterred by language barriers but instead will try to study it in order to discover the author's thoughts, or they will seek out those who will translate and publish the article. Nevertheless, even the most renowned scholar will find it necessary to communicate his discoveries to the most widely circulated German *Beitrag* or *Zentralblatt* in order to save time and win over the experts. Beginners who have not yet gained recognition in the world of scholarship would be wise to request at the outset publication in the great foreign journals, and to write in French, English, or German, or to have his work so translated. The new data will become known rapidly to specialists in this way, and if it has positive value the author will have the pleasant surprise of seeing it confirmed and approved by the great international authorities. People who are inspired by a narrow-minded, ruinous patriotism and insist on writing exclusively for the Spanish journals—which are little read or totally unknown in countries renowned for their intellectual achievements—condemn themselves to being ignored even in their own nation. Because they will always lack the *exequatur* of the great names of Europe, none of their countrymen, and least of all their own colleagues, will dare take them seriously or value their true worth.

Because the judgment of scientific authorities abroad is critical in determining the future of the young investigator, he will reflect thoughtfully before submitting his first work to them. He will assure himself of the truth and originality of the data by way of painstaking bibliographic research and, better yet, through consultation with some famous specialist. And he must not forget that the right to make mistakes is tolerated only in the famous.

9 The Investigator as Teacher

Once the constructive period has been reached, and the difficulties of scientific work have been overcome, we imagine our young investigator will have the maturity and strength necessary for spiritual development. The noble career has been pursued successfully, the ideal so anxiously sought has been attained. Transformed into an international authority, the master is cited with praise in the foreign journals, and the originality and importance of his concepts assure him a page of honor in the golden annals of science.

Under such favorable circumstances, the scholar may adopt one of two positions: he can pursue his laboratory undertakings alone with complete concentration, sentencing himself to educational sterility; or he can make others participants in his methods of study, advancing the cause of worthy vocations and establishing himself as the renowned head of a school.

The advantages of both paths are clear. Solitary work definitely satisfies the ego and is accompanied by a tempting tranquility. One obeys the law of minimum effort, directing attention exclusively to personal research, and lives in a quiet atmosphere of approval and esteem—although great respect and enthusiasm are lacking (which, however, may be a great advantage). On the other hand, one is not bothered by imitators and rivals. But—the paternal instinct within the scientist makes him extremely restless when adopting this comfortable position. "What will become of my work," he asks himself, "when I am old and lack the energy to protect it? Who will defend the priority of my discoveries if adversaries or unscrupulous successors happen to appropriate them or commit oversights and injustices in judging them?"

Even considering the situation from a selfish point of view—with a sound and clear-sighted ego—it is important for the scholar to advance his spiritual development. The task is unavoidably painful. The master's activity is divided between the parallel currents of the laboratory and of teaching. Thus, anxiety will grow, but opportunities for happiness will blossom as well. By encouraging lofty inclinations, the master's delight in intellectual paternity will be fulfilled, and he will feel the noble pride of having fulfilled his double mission of teacher and patriot honorably. Now he will not approach the end of his life alone and sad. Instead, he will be surrounded during his decline by a group of enthusiastic disciples capable of understanding the work of the master, and, insofar as possible, of making it a shining and lasting memorial.

Posterity has always been generous with the founders of schools. Even their errors are pardoned or kindly explained away if they were able to develop the spirit of understanding and correcting mistakes. In contrast, he who renounces the sowing of new ideas pronounces himself an egotist or misanthrope. There will be a sense that he pampered his pride instead of laboring for humanity. And if he has really outstanding talent, he will seem a bit pathological, an excrescence apart from his race and thus hardly praised. He is like a shooting star fallen from the heavens, shining brightly for a moment but incapable of imparting its ephemeral brilliance to anything.

Besides lending great worth to the scholar's life, leaving spiritual progeny has undeniable social value, and is ennobling work. Countries like Spain, where the production of scientific achievements is so wretched and sporadic, desperately need this type of effort.

Unhappy is the occasional genius who appears in these countries and is extinguished without offspring! The severe competition among hundreds of foreign laboratories and schools, the crushing avalanche of reprints and books hotly struggling for today's favor, the iconoclastic tendencies of university students eager to *arrive* and to establish and maintain their own personalities, the almost total ignorance among scientists of languages spoken in the backward countries, and above all, the savage *chauvinism* reigning in Germany, France, and England—in unfortunate complicity with the national apathy of Spain—have had the saddest of consequences for the proud recluse in the well-known ivory tower. Many of his discoveries inevitably will be attributed to foreign collaborators who may not be overly scrupulous in their citations, and to their even less scrupulous pupils. And data that seemed trivial at the time of publication (and did not merit the honor of translation), and that have a way of increasing their value as time goes on, will be buried in the dust of our own libraries. While critics and scholars abound in the fields of literature and history, which are recreational and charming arts, in the austere discipline of science the defender must be wise as well as learned, and the wise are not abundant in countries without sufficient culture!

Therefore, it is important for these backward countries to obtain from those who promote science the maximum teaching efficiency possible, compensating as much as possible for losses as their workload increases.

But how does one train disciples able to carry on the work, or better yet encourage geniuses that will be able to surpass the master, identifying new roads of investigation?

And having come this far, another important question arises. How is an irresistible affinity for science created?

Although Fouillée, Ribot, Bernheim, Lévy, and many others have justifiably said that all ideas accepted by the brain tend to be converted into action, there is no doubt that most people lack the ability to transform scientific ideas or knowledge into the *act* of confirming them, or into the act of expanding the horizons of knowledge through personal effort.

In our view, the young man's will operates through the motivation of anticipating the pleasures associated with any intellectual triumph. The feeling of self-esteem grows when respect comes from the learned. And the opposite is also true: if we are scorned we begin to scorn ourselves. This fact generates the need (which, unfortunately, is too often overlooked completely) for teachers to remind their pupils often of the great joy, the supreme satisfaction, produced by wresting secrets from the unknown, and by perpetuating one's own name through original and useful ideas. More often than not, these suggestions are made by example rather than word.

It is well known that youth show their respect for famous men by imitating them. Therefore, it would be a worthwhile contribution to the education of the will if each and every teacher would recount with genuine affection, and with the deliberate intent of suggestion, the anecdotal and more formal biographies of the scientists who have distinguished themselves most in the development of the student's chosen field. This would foster something of the spirit intended by the following authors: Comte, with his veneration of great men; Carlyle in modern times, with his book of heroes; Emerson, with his enthusiastic apologias on representative men or supermen—to whom we owe all of the progress and advantages of civilization; and finally, Ostwald with his inspiring book, *Great Men*.

What signs identify creative talent and an irrevocable calling for scientific research?

This serious and fundamentally important question has been discussed at length by deep thinkers and noted teachers, without coming to any real conclusions. The problem is even more difficult when taking into account the fact that it is not enough to find clear-sighted and capable minds for laboratory research; they must also be genuine converts to the worship of original data.

Are future scientists—the goal of our educational vigilance—found by chance among the most serious students who work diligently, those who win prizes and competitions?

Sometimes, but not always. If the rule were infallible, the teacher's work would be easy. He could simply focus his efforts on the outstanding prizewinners among the degree candidates, and on those at the top of the list in professional competitions. But reality often takes pleasure in laughing at predictions and in blasting hopes. Fervidly virtuous and earnest young men often prove to be extremely egotistical, and one also finds with distressing frequency examples of the most brilliant young men whose acutely practical minds are hatching sophisticated financiers. They study and work less for the love of science than for a belief that knowledge constitutes good business, and that a good reputation won in school is valued highly in the professional marketplace and in academic circles.

If the reader smiles with disbelief at this observation, let him reflect a moment on the fate of his most brilliant classmates—the prodigies of memory and action in whom the teacher indulged everything, to whom all preferences were directed. He must acknowledge with chagrin that while the majority have reached a level of social comfort (and were thus correct in their calculations), very few if any have ascended the heights of knowledge or distinguished themselves by unselfish and fruitful social or industrial accomplishments. And this occurs all the more frequently because a fair number of the most proficient students have a rather weak personality. They tend to be meek and disciplined, and to lack initiative. Having accepted a course of study through blind obedience to parents and advisors, they often end their careers overcome by weakness and fatigue. Who has not heard those forced to read a textbook exclaim at the end of the course: "Goodbye Horace, whom I hated so much!"

The clear-sighted teacher will find another type of student much more worthy of attention. They are somewhat headstrong, contemptuous of first place prizes, and immune to the inducements of vanity. They are endowed with an abundance of restless imagination and spend their extra energy pursuing literature, art, philosophy, and all the various recreations of mind and body. To a distant observer, it would appear as though they are spreading their energy too thin, whereas in reality they are channeling and strengthening it. These versatile young men are generous souls—poets at times, but always romantics—and they have two essential qualities that the master can turn to excellent advantage. They scorn material gain and high academic rank, and their noble minds are captivated by lofty ideals. In contrast to the others, they really begin to study after leaving the classroom. One commonly sees them—exhausted from laboring without progress and without clear bearings—appearing in the laboratory and begging for technical advice and a problem to work on. A few of them eventually succeed in orienting themselves and achieving success.

However, the traits just mentioned are not foolproof earmarks of success. There are many inconsistencies and defections among those who display an abundance of these traits, which early on represent potential energy that does not always materialize. Deceived by appearances, the master runs the risk of training laboratory dilettantes or brilliant minds incapable of thorough and persevering work.

The diagnosis of a scientific calling is therefore difficult. One must use finer distinguishing signs to cull the genuine from the counterfeit.

Ostwald has dealt with more or less the same problem in his admirable book, *Great Men.* He believes that especially gifted students may be recognized by the fact that they never appear satisfied with what ordinary instruction offers them: "In terms of depth and range covered, ordinary instruction is directed toward the average student. When a pupil has great talent, he will see at once that the science being taught is inadequate quantitatively, and above all qualitatively, and he will demand more." Then he adds: "The most important quality of the scholar is originality, that is, the ability to picture something beyond what is taught. Precision in one's work, self-criticism, conscientiousness, knowledge, and skill are also necessary, but all can be acquired later through suitable education."

These observations of Ostwald are judicious and generally correct. However, for the master to benefit from them, he must be in friendly contact with his students. In his laboratory discussions he should treat them like colleagues working on a common goal, encouraging frankness and spontaneous expression. In doing so, the master finds opportunities to study the character of his pupils, as well as to gauge their vigor and firmness. Even so, Ostwald's rule fails occasionally. The clever lad who is dissatisfied with textbook descriptions and scientific theories may possess a high-minded character and keen intellect, and yet lack perseverance and discipline. More often still, the young investigator is overly timid. His respect for the master, and a rather winning modesty, curb his desire to request clarification of his doubts about certain theories, or request approval for experiments to test new hypotheses. Under these circumstances, he may not be noticed by the professor, or may not be encouraged enough by him. Unfortunately, his reserve may have been mistaken for inherent limitations.

We believe the following guidelines for psychological diagnosis are somewhat more dependable, although they are certainly not infallible. They combine subjective and objective signs.

Subjectively, the young man suited for investigation can be spotted at once by the following traits. He has an ardent yet sane patriotism. Instead of sharing the naive optimism of certain patriots—basically *professional patriots* who mention four or five famous Spanish names in trying to demonstrate the positive contributions of their country to the achievement of universal culture—our young man feels a profound discontent with the poverty and paltriness of such contributions. Faced with the severe yet well-founded judgment passed by foreign criticism on the intellectual sterility of our scholars and philosophers, he does not respond with laments or boastful promises. Instead, he sharpens his weapons, and resolves to dedicate his energy to the universal struggle with nature. Our potential scholar is also distinguished by a rigid worship of truth, and by a sound and genuine skepticism. He is ambitious but with noble and worthy goals. He is eager to remove himself from the atmosphere of the ordinary and link his name to some great enterprise.

Objectively, candidates for the title of savant convince everyone that they have the promising traits just mentioned. Without proof that the novice is industrious, and has the ability to manipulate data, we run the risk of cultivating just another boastful reformer—as willing to point out the road as he is unwilling to cross the gulf in front of it. But if the young man thoroughly enjoys laboratory work and has boundless energy; and if above all (and this is the key objective sign we have alluded to) we discover that, at the cost of painful sacrifice, with economies stolen from recreation and amusement, he has established a small laboratory where he works eagerly to acquire technical mastery and to confirm personally the discoveries of eminent scholars, then the professor must resolutely intervene by helping and shielding him. *True vocation always consists of a special activity for which the young man sacrifices time and money, scorning the distractions of our age.*

Sometimes this trait can be misleading, of course, even though sincere and enthusiastic. Vocation does not imply aptitude, nor does aptitude necessarily guarantee success. The latter has a complex genesis because complementary elements enter into it. They include the shrewdness to track down rich lodes of knowledge, a gift for assimilating new ideas, a sure and penetrating critical sense, a good bibliographic and methodological orientation, and even a certain philosophical turn of mind. However, most of these complementary traits may be acquired later. Something must remain for daily contact with the master and for the transforming power of imitation.

To summarize: The future scientist is typically an ardent patriot who is eager to bring honor to himself and to his country, captivated by originality, indifferent to material gain and ordinary pleasures, inclined more toward action than words, and an untiring reader. In short, he is capable of all sorts of sacrifices to achieve the noble dream of giving his own name to some new star in the firmament of knowledge.

A Critical Optimism

We have already stated that a master worthy of the name must always convey to his pupils the idea that science is in a perpetual state of flux, that it progresses and grows continuously, and that we can all contribute a grain of sand to the imposing monument of progress if we truly resolve to do so.

A similar attitude is involved, of course, in national optimism—a strong faith in the abilities and destiny of the race. This optimism should not be blind, of course, but cautious and farsighted instead. The conscientious leader should be fully aware of our national lack of culture and our scientific poverty, rather than entertaining the regrettable and smug conceit of many politicians and more than a few pompous classroom orators. He will always bear in mind that for centuries Spain has been in debt to civilization, and that if she persists in such shameful neglect of duty, Europe will lose patience and eventually ostracize her. Criticize but work; censure and if necessary reprove the indolent, but without looking backward; and keep the hands firmly on the plow.

Some of those who have shared in this patriotic optimism (called by Godó "paradoxical optimism," although "critical optimism" would be a better phrase) include the great Costa, whose apostrophic invectives cracked like lashes against the backs of the laggards and the faces of the unpatriotic; and in modern times the highly accomplished writer and thinker Ortega y Gasset, who suggests that for Spain to advance ethically and culturally we must recognize fully our spiritual poverty and our political and administrative corruption.

How to Guide the Inexperienced Investigator

Once the intellectual family is selected, it must be educated and trained for hard work. It would be puerile and reckless to compete in scientific contests flavored with serious international competition without strong and adequate preparation.

The master has the responsibility of shortening this preparation, orienting the pupil, showing him roads open to investigation, guiding him in literature searches, and finally, suggesting to him how to acquire all of the accessory knowledge and abilities that are useful: drawing, photomicrography, languages, and the art of describing properly and accurately. It is important to instill in him the determination to complete his education in these respects as soon as possible so as to avoid the ongoing humiliation of collaboration, which obviously can't last forever.

With the neophyte's mental powers strengthened in these ways, the teacher will try to put them to the test by suggesting a topic within his grasp—one that does not require unusual or lengthy effort and which, if at all possible, is an extension of the master's fundamental work.

It is well known that youth have a tendency to attack the major problems and begin their careers with a major piece of work. It is necessary to restrain such ambition, which might easily lead to discouraging failures, and make the beginner see the advantages of starting with the minor problems. He runs little risk of committing mistakes with them, and when he does there is no chance of ridicule. Later on, there will be opportunity to carry out the great work he dreams of, when technical aptitude and greater understanding are developed.

When the beginner can finally walk alone, attempts should be made to infuse him with an appreciation of originality. Thus, new ideas should be allowed to come to him entirely spontaneously, whether or not they agree with the theories of the school. The greatest honor that can come to the master does not lie in molding pupils to follow him, but in producing scholars who will surpass him. The highest ideal would be to create absolutely new spirits—if possible, unique additions to the machine of progress. The creation of pliable and interchangeable instruments highlights the master's greater preoccupation with himself than with his country and with science.

There in no need to caution the master that in his books and monographs he should always render sincere justice to the student, citing his work scrupulously and even dwelling on it with encouraging pleasure. More through consideration for his intellectual offspring than modesty, he will remain silent about his own contributions. The beginning scholar's prestige will grow in this way, and his work will be accepted quickly and with friendly sympathy abroad.

Many scholars are accustomed to adding their own names to the first papers of their students. This highlights the collaboration, but while certainly appropriate it is hardly generous. Unless this initial work is due almost entirely to the master's personal supervision, it would be preferable to free students from the somewhat humiliating suggestion that the inspiration was not their own. Young investigators may then relish the exquisite experience of originality. Once experienced, it is rare indeed not to develop a liking for creativity, and not make every effort to foster it.

It also seems unnecessary to recommend that professors avoid taking undue advantage of a docile student's energy under the pretext of directing and preparing him. This abuse reveals an offensive egotism and thrives in some schools abroad, where the novice in certain professions pays for his schooling with the exploitation of an apprenticeship. How often monumental work reflects not the author's productivity but the discretion and modesty of youthful collaborators who are content with the distant hope of someday receiving their intellectual mentor's backing for suitable employment.

Scholars occasionally descend to such reprehensible exploitation because of hard times, but more often in eager pursuit of honors and awards, which are incompatible with a peaceful life and with thorough and persevering work. Having arrived with honor, one must fall with honor, and personal merit should be enough for anyone. The master is repaid amply by the satisfaction of having awakened latent abilities and shaped creative minds. If weakened senses and failing will deprive the aging man of the vigor necessary for research work, he should resolutely abandon his active professorship. One can only teach effectively those things that one actually does, and he who is not doing research cannot instruct others how to do it. Gracián has pointed out that the ultimate in discretion is "to know when to stop." Even though it may be painful, we must at a certain age forsake teaching before teaching forsakes us.

Nevertheless, the veteran professor still has a valuable mission to accomplish. When his weak hands can no longer swing the miner's pick, he can occupy himself with refining the material recovered by others.[1] Let him write the history or philosophy of science in the peace of his retirement. No one can elaborate better on it than he who has lived through its events and experienced firsthand its difficult theoretical issues.

Note

1. Nature has been merciful to the aged, granting the brain the sublime privilege of resisting more than any other organ the inexorable process of degeneration.